MISS BUMBY'S MISSION

JM Laird

MISS BUMBY'S MISSION
JM Laird

ISBN: 978-1-914934-91-9

© JM Laird 2025. JM Laird asserts their right to be known as the author of this work.

Published 2025 by Northern Bee Books, Scout Bottom Farm, Mytholmroyd, Hebden Bridge HX7 5JS (UK). 01422 882751.

Cover photo by EA Hansen, *Bonnet for Mary Bumby.*

Book design by www.SiPat.co.uk.
Typeset in Baskerville and Minion Pro.

All rights reserved. No part of this publication may be reproduced, stored or transmitted in any form or by any means electronically or mechanically, by photocopying, recording, scanning or otherwise, without the permission of the copyright owners.

MISS BUMBY'S MISSION

JM Laird

CHAPTER 1

Thousands of black insects hurtle around my head. Their tiny translucent wings create a hum in the air. Most avoid touching me, but occasionally a bee bounces off my veil.

'Settle down, girls,' I say, blowing one of the small bodies from my veil. 'I know the chafing dish has gone out. But I'm not here to harm you.'

Talking to my bees is a habit I learnt from my old teacher. I can still hear his voice in my head.

'It don't matter that the beasties don't understand you, Mary. 'Tis all bout how you say it and keeping yerself calm. It helps slow you down, gives you time to plan what you want to do.'

I pick up my skirts with a gloved hand and move away from the skeps. It's a glorious spring day, with a cloudless sky and only a gentle breeze. I'm delighted to be back in the countryside, away from Birmingham's smoke and din. Back out with my bees. Today seems a likely swarm day, when many of the bees will take off from their hive and form another colony. I want to be nearby in case they do, so that I can recapture them. I've brought another skep, made of straw and bound with bramble, to house them in, rather than losing them altogether.

I look around, noting the dandelions dotting the fields. In the distance is a haze of pink and green, the flowers of the heath. I must remember to explore these fields later in the year when the bilberries are ripe and I can gather some of the blue-black fruit for jam. Back towards the road, the cab man lies on the grass with his hat over his face, while his horse grazes contentedly nearby.

I collect some more of the cow dung I've brought and place it in the chafing dish. Then I strike a light with the Bell & Black lucifers I've hidden from Cook and get the dung smoking. It will calm the bees if I leave it by their entrance.

I don't usually tell people that I keep bees. It is an odd occupation, meant for vicars and monks so that they can get wax candles for their churches. My own church, the Thirsk Wesleyan Chapel, benefited from my beeswax before I moved to Birmingham with my brother. Reverend Dakin was saddened to see me leave. My candles, and my lovely alto voice, and my cheery nature, he'd said.

I like being a skeppist because it is an excuse to be outside, instead of sitting in a dim room doing needlework. I also enjoy working with the bees. I love their busy lives, constantly on the move for the betterment of their sisters. I've learned that if I am gentle, they won't bother me.

And my brother, John, adores my honey. Every morning, without fail, he has toast with a liberal spreading of my golden offering, courtesy of my bees. How can I fail to adore them for providing such a treat?

Now that the chafing dish is smouldering, I move back to the skeps. They are sheltered in the cavity of a stone wall. I heft each one to check their weight and gauge how many bees they contain. It looks as if it is going to be a good season.

'Come on, girls. Let me see what you've been doing over winter.'

I almost choke when my brother mentions moving to the other side of the world.

'After much heart-searching, Mary Anna, I've decided that I must venture forth into the wider world,' John says, reaching for another slice of bread and turning to the fire to toast it. 'I feel called to bring the word of the Lord to those souls who have yet to hear of Him.'

I lower my cup of tea with shaking hands. John is a minister in the Wesleyan Church, and the Lord's work is his life. 'I've heard that India is full of pestilence, and the climate is rather hot. I know that sounds appealing after the winters we've had recently, but do you think your health will cope?'

I look more closely at him. He is still too thin for my liking, perhaps because he strictly controls what he eats and is fiercely determined not to sleep overmuch. He's also inclined to cough, especially here in Birmingham where the nights are thick with that sluggish grey fog. It worries me; I don't want to lose another sibling.

I ring the bell for Nancy to come and refresh the teapot.

My life is bound to John's. I live with him as his housekeeper, dedicated to easing his life so that he can concentrate on matters of the spirit. It hasn't always been this way. In my childhood my constant companion was my younger sister Jane, while we let our older brother pursue his bookish interests. But then Jane died. And when Mother was also taken, a few years later, I transferred all my affection and attention to John.

He spoons a dollop of honey onto his toast. Honey from my hives. How he loves my honey. 'No, not India, Mary Anna. The islands of the South Pacific, particularly New Zealand. The news I've heard is that the country is beset with whalers, sealers, and merchantmen. There are few missionaries there and the natives are not learning anything of our Lord. It is my duty to go there and introduce them to Him.'

I struggle to remember what I've heard of New Zealand. New South Wales is more prominent in stories of the colonies, what with all the felons being shipped there to work out their sentences. And there are just so many other new places in the world that our countrymen are conquering and exploring, it's hard to keep abreast.

I've never been on a boat or experienced the ocean. The furthest from home I've ventured is here, Birmingham. We were brought up far from the sea in the small town of Thirsk, in Yorkshire. Father hoped that John would follow in his stead as a butcher, but from early on we were both influenced by our mother's faith. The result is that John is now an ordained minister.

'I'll come with you if you get the commission,' I offer impulsively. It seems a far-fetched idea. I can't believe he is serious about travelling so far from home. He's never visited the Continent, or even Scotland, a journey of not much more than one hundred miles. Surely, he'll change his mind. He isn't that sort of man, to put himself in danger.

John stares at me. 'I couldn't ask that of you,' he says. 'It's such a long way. And New Zealand is still an uncivilised place.'

'I promised Ma that I would look after you,' I say. 'And Father is well cared for, with his new wife and all his cousins. There's nothing here to hold me.'

Indeed, since our mother died, I've dedicated my life to John. It's what she wanted, and I feel it is my honour-bound duty to both my family and my Lord, enabling John to devote his attention to his work.

'You've many friends among our congregation. Mrs Hyde and Miss Sells, especially. They'd take you in, I'm sure.'

I hate the thought of being a charity case. Although I'm beginning to regret the rashness of my offer, John won't persuade me otherwise.

'John, if I stay, I'll be a spinster without a chaperone. I'm twenty-six and unlikely to find a husband now. Besides, what was the point of Ma teaching me to read and write if I don't use that for the glory of God? It's clear to me that His purpose is for me to do my duty and follow where you go.'

It would have been different if Jane had lived. We could have been spinsters together, two sisters doing good deeds in the community.

I don't say it, but we both know that the other option would be for me to become a governess, teaching the bratty children of the privileged to read, write and behave. While I like children, I have no desire to live as a glorified servant in an ungodly house.

John brushes the crumbs from his fingertips and reaches across for my hands. 'Mary Anna, you are a good and kind sister,' he says. 'I would welcome your company and assistance.' He withdraws his hands to cradle his teacup. 'Besides which, it is far from decided. I must still make my case to the Conference.'

Exactly, I think. They'll make him see sense.

I continue to feel anxious about John's decision and my reckless offer. It's occupying my mind when my friend Miss Emma Sells visits the next day. Her blonde hair peeks out of her bonnet, so fair in contrast to my own dark curls. Emma would be pretty, I believe, if it weren't for her overlarge nose. Not that I don't have my own faults, I remind myself. I must try to be more charitable.

'I've just heard that old King William has died,' Emma says breathlessly. 'Isn't it exciting?'

I frown at her. 'The Lord keep his soul and grant him peace,' I say. 'It must be only ten years since he came to the throne. Remind me, does he have any surviving children?' An old king is not such fun to gossip about, but now we might have a new king.

'Not any from the right side of the sheets, unless the old Queen is expecting. So, we are to have his niece as Queen,' Emma replies, her eyes glowing with excitement. 'She's only eighteen, and called Drina.'

'Drina? That's not a very regal name. When will they have the coronation?'

'I hear she's to be known as Queen Victoria. There's lots of planning for a coronation. Getting heads of state from other countries invited, and the clothes and carriages. It won't be for another year.' Emma's family visits London regularly and she always knows the latest happenings.

'It must be over a hundred years since we've had a queen ruling alone,' I muse. 'And she's so young. I wonder how she'll cope.'

'Yes, it's hard to see how a woman can hold on to our remaining colonies. Father says we mustn't repeat the mistakes we made in America.' Emma's father has links with the East India Company and is inclined to promote the concept of British superiority. He didn't approve of the end of slavery.

'Speaking of the colonies, my brother is considering missionary work in the Southern Isles,' I tell her. At one point I'd entertained thoughts that Emma might make a good bride for John, although she is getting rather old for marriage. The Anglo-Burmese War, and the loss of so many English soldiers, has made it difficult for girls like us to find marriageable young men. But I've come to realise that Emma is much too interested in clothes and the right people to suit John, and she's certainly not devout enough.

'How frightful!' Emma says. 'I've heard that the natives are cannibals, using the whaler's trypots to cook their foes.'

I'm also concerned about the stories I've heard of natives eating their enemies, but I've resolved to be brave.

'I'm sure those reports are exaggerated,' I tell Emma.

'You've been reading overly many penny dreadfuls.'

'But Mary, if your brother leaves to become a missionary, what will you do?' Emma asks.

'Why, I will accompany him, of course,' I say, looking forward to Emma's reaction.

'No! You can't!' Emma covers her mouth with a gloved hand, her eyes wide.

'But I must. It is my duty, both to my brother and to the Lord,' I say. Saying it seems to make it more true.

'Oh, Mary, you are so good. To sacrifice yourself in this way. I can hardly believe it,' Emma says, fanning herself with her hand.

I struggle to hide my amusement. I have no doubt that our acquaintances will be regaled with this news within days, and the new queen will be relegated to second-best.

'Nonsense,' I say. 'It will be a great adventure. Why should men have all the fun, after all?'

'Fun?' Emma squeaks. 'But how will you . . .?' She struggles to find an example of how ghastly my life will become. She looks down at the teacup in her hand. 'How will you get your tea?'

'I expect that supply ships will bring it from here, or maybe directly from China,' I say. 'Or maybe we'll change to drinking coffee.'

'Coffee! Ugh. You know, Father says they're hoping to grow tea in India,' Emma says, her voice lowering a tone. 'Except I'm not to tell anyone. You'll keep it a secret, won't you? Imagine, then it will be a truly English drink.'

'Oh, Emma, I am going to miss you,' I say, unable to conceal my mirth any longer. 'Of course, I won't tell anyone about Indian tea. Now, tell me what you saw last time you were in London.'

At first it isn't certain that we will be allowed to go to New Zealand. The Conference is keen for John to remain in England and continue his pastorate. It's an unsettling time for us both, not knowing what the future holds. John remains adamant that this is his calling, despite the counsel of some of his fellow ministers.

'They don't understand,' he complains. 'They keep talking about all the people here who still need to be brought to the Lord. And now this weird religious group from America has landed at Liverpool and started making converts. They call themselves the Church of Jesus Christ of the Latter-Day Saints, would you believe? As if new saints could be made. It's bordering on Catholicism.'

'It's true that there's a lot of work still to be done here,' I say. 'The new Poor Law makes life difficult for many. They could do with some good tidings, such as the presence of our Lord in their lives.'

'But the very point of Mr Wesley's teachings was that *all* people should be given the chance to be saved. These natives also deserve to hear the Word.'

I know how determined John can be. I've been reading anything I can find about New Zealand, although it isn't always reassuring. In particular, the old reports of the *Boyd* massacre, when the natives killed and ate about sixty people on a stranded ship, make me shudder.

'That was many years ago,' John says when I mention it. 'It shows how much we need to be there to lead the natives to civilisation.'

'Yes, but then what about the case of Mrs Jacky Guard?' I ask. 'That was only a few years ago.' Betty Guard had been held captive with her two young children after their ship ran aground. Her husband and other men escaped and went to seek help and came back from Sydney with soldiers.

'She was returned safely to her husband,' John says.

'Yes, but she reported that the natives ate some of the sailors they'd killed.' And who knows what other liberties those savages may have taken with her, but it isn't polite to mention that.

'The natives don't know any better, Mary Anna. We must teach them our ways. Indeed, our people didn't behave much better. The Parliamentary enquiry showed that our soldiers opened fire on unarmed natives. No wonder they don't trust us. We missionaries need to be there to help protect them. Especially now that we have these companies setting up to send out settlers.'

I've seen advertisements in the *Times*, promising land for emigrants and work for labourers in towns that aren't even built yet. The New Zealand Company, headed by the Wakefield brothers, is particularly energetic in its efforts. But it isn't the only company promising riches for the embattled labourers.

'I hear there are only a thousand white people living in the whole country,' I tell John. 'Will we be in one of these new towns?' There are more people than that living in Thirsk, and that's considered a small town.

'We'll go where we're needed,' John replies, 'where the Conference directs me. But this is my decision. Yours doesn't have to be the same.'

I swallow my misgivings. 'I know,' I say. 'I just want to be sure you know the situation we're heading towards.' I think about Emma's concern over supplies of tea. I'll have to ensure I have adequate provisions to last, sealed and secure to survive the sea journey. What will life be like without merchants and carters, I wonder. And what can I give the natives in exchange for not being eaten?

Although I'm still uncertain whether we'll get to go on our mission, I'm determined to learn some practical skills.

'You look as if you're spending the day in the country,' John says one morning.

I'm surprised he's noticed. He never usually comments on anything I wear. Today, I'm wearing one of my oldest dresses and a stained apron and have my third-best gloves ready.

'I'm visiting the skep-maker,' I say. 'I want to understand how they are made.'

'For your bees?' John asks. 'Can't you just order new ones when you need them?'

'Of course. When we're here. But if we go to New Zealand, I need to know how to replace them.'

'Well, that raises the question of whether they even have honeybees in New Zealand, doesn't it?' John asks. 'Have you been able to ascertain that?'

'I can't tell,' I say. 'Nothing I've read says anything about them. Maybe they are too domestic to mention. To be sure, though, I thought I'd take a couple of my hives. The ones with the quietest bees.'

'All that way? Is that a good idea? It seems a lot of effort when they might be there already.'

John has never paid much attention to my little pastime. I'm surprised that he doesn't consider harvesting the honey to be the extent of what I do.

'Still, it would be wonderful if you did take some bees. I've been trying to reconcile myself to the idea of not having any more of your honey, but I can't imagine tea and toast without it,' he continues.

As I had thought. He doesn't understand that the bees are my charges, with their welfare my chief concern. I think of another reason to persuade him.

'I thought the wax might be useful for candles for the chapel, as well,' I say.

'Oh yes, I can't abide having to use tallow,' John says. 'They stink the chapel out. Whereas wax candles represent Christ's sacrifice and are sweet-smelling and far less spluttery.'

I think I have won the argument.

I spend the day with the skep-maker in his hut. In Thirsk, skep-making was another task for the basket-maker, but here in Birmingham the greater number of beekeepers from the surrounding countryside mean this man can specialise. I haven't paid much attention to how my skeps are constructed before, apart from ascertaining that they are of sufficient size and strength for the vast numbers of bees who make their home inside the cosy basket.

The skep-maker is surrounded by straw, as if he is preparing to thatch a roof. In fact, the place smells rather like a cottage after sustained rain – that musty vegetative smell. He sits me on a stool in front of him, so that I can watch. Then he grabs up a handful of straw and feeds it into a length of hollow cow bone which compresses the strands into a loose rope. He threads an awl through his previous work and binds the straw bundle with a strip of bramble cane poking through the awl, while twisting the straw bundle to curve it.

'Here, Mistress, have a go,' he says, after twenty minutes of demonstrating the technique. 'Nothing like hands working to get a feel of it.'

'I don't think I was paying enough attention,' I say. We exchange stools and I try to remember what he'd been doing.

'That's right, Mistress. Grab a handful of that there straw. Now twist.'

'It's quite hard to do, isn't it?' I say, surprised. 'And wet.'

'Put this here apron on, Mistress. 'Tis leather, it is, so it might keep you drier. And your poor gloves.'

'Yes, they're quite sodden. Never mind. Why is the straw wet?'

'To make it bendy, instead a breaking. Yes, that's the way. Twist that straw right tight.'

My best attempts only produce a loose and distorted row of binding, and I feel a new appreciation for the craftsmanship of this man. I hand the mangled skep back to him, and he tosses it aside. He'll have to start again, I realise. I'd thought it would be a skill I could easily learn, but it is more difficult than he made it look. It is certainly very different to needlework, which should be useful for mending and making new clothes but doesn't require the same strength.

I think about my straw bonnet. How will I replace that once it wears out? And boots? How will I cope without all these skilled people in my new country? Is there even straw to work with?

It isn't going to be easy; I can see that. I knead the tight muscles in my forearms, hoping that my resilience will be strong enough to endure the trials ahead. I understand that I'm going to be tested, many times over.

'An' how will you be cooking your supper, then?' Cook is asking. 'One of those darkies will be doing it, with foreign muck I suppose.' She bangs some pots onto the stove as if in protest.

'Well, I imagine they cook food for themselves,' I say. 'It can't be difficult to ask them to cook for us as well. Besides, Mother taught me to cook.'

'I don't think the young master should be taking you to such a place,' Cook says. 'It's not right for a young lady such as yerself to be among heathens.'

'But I want to go,' I protest. 'I need to support my brother. And it's only right that I help bring His Word to the people there.'

'Well, I can't say I won't miss you, the both o' you. But afore you go, I'll learn you some of the things that might keep you from hungering. First thing you needs to know is how to set a fire.'

'Cook!' I grin. 'I know how to light a fireplace. The kindling, then bigger pieces, then coal.'

'Ah, but what will ye use to strike a flame?'

'Lucifers, of course. And then I'll bank it down at night so it can come away in the morning.'

'Those strike-anywheres? New-fangled things; I don't trust 'em. My cousin's husband's sister had her house burn down, and they reckon it was rats eating the wax. And what if you gets them wet, in a storm or such? No, there's nothing such as a good old-fashioned tinderbox, if you're asking me. They lasts for ages as well, so you don't need to keep buying em.'

'You're right, Cook. These are things I should learn about. I'll become your meek little student and you can teach me what I need to know. What shall we study?' I ask, suppressing my smile. I should have remembered Cook's dislike of the self-igniting wax tapers, and indeed, anything new.

'Well now, you should know how to joint your rabbit and clean your pigeon, how to make suet pastry and preserve fruit. How to bake bread, keeping the barm going so that the bread does rise.'

'I'm not sure that they have rabbits there,' I say. I've read that the country has few animals, apart from those brought in by the English.

'No rabbits? Then what does the poor people eat?' Cook asks. 'What about pigeons?'

'I don't know. I guess they have birds of some sort.'

'And I knows that you're handy with your needle, but you needs to learn how to darn and patch,' Cook says. 'And maybe I could learn Mr Bumby some of the tinker's trade.'

'Teach me. My brother will be busy with his mission.'

'Eeh, rather thee than me, lass,' Cook says, wiping her hands down her pinny. 'Tha' knows I love you dearly, but I'd no leave my country for love nor brass.' It's a strange thing for her to say, considering she's come with us from Yorkshire to Birmingham. She's already told me that she'll be returning 'home' when we leave.

Truth be told, I'd rather not leave England either. There are so many things I don't know about New Zealand. Only a few missionaries have returned and of those, not many had taken wives or housekeepers. It sounds as if the Englishmen there tend to be of a rough class, illiterate and coarse speaking. I can't imagine how I will fare.

Still, it is my duty, as I keep reminding myself. Both to my brother, and to my Church. If my sex has ever rankled with me, limiting the choices I have compared to John, then I cannot shirk this opportunity to show I am equal to the task.

CHAPTER 2

When my bees start collecting the first pollen from the dandelions, John leaves to attend the Wesleyan Conference in Bristol with his great friend Mr Waterhouse. I wait impatiently to see whether the church leaders will accept John as a missionary. He is so set on it that even though I'll be secretly pleased if we remain in England, I can't wish it for him.

When he arrives home from Bristol, it's in the company of his fellow minister Mr Barrett. I don't need to ask what the outcome is. They are both full of smiles and chatter and back-slapping bonhomie.

'Miss Bumby, you are now looking at the new Superintendent of the Wesleyan Mission in New Zealand,' Mr Barrett says, pointing to John as we sit at dinner.

'Brother, they could hardly fail to appoint you,' I say. 'It is well deserved, after all your work here. I know you're young, not quite thirty, but you've certainly shown your dedication.'

'Well, I wasn't at all sure about it,' John says. 'I felt I had to argue my case quite strongly. Of course, Mr Waterhouse supported my application, and that helped.'

'Reverend Bumby's appointment was met with great approbation, you know,' Mr Barrett says, helping himself to more of the stew that I've prepared with the help of Cook.

John blushes.

'Your brother is so modest,' Mr Barrett says, noticing John's colour. 'In fact, he was so overcome with emotion that he struggled to address the Conference.'

'I meant what I said,' John mutters. He turns to me. 'I told them it might be the last time we should meet on Earth.'

I put down my cutlery as a chill works its way down my back. 'Oh John, what a thing to say!'

Mr Barrett nods. 'I freely admit I wasn't the only one who wept at his words.'

Apart from the cannibalism, it hasn't occurred to me that we might never return to England. I thought it would be a matter of years in exile before we returned, not a life-time sentence. Now I feel as if I'm about to be punished like a convict, sent to spend my years away from home. I struggle to maintain my composure and seek to divert the conversation. I turn to Mr Barrett. 'You've been a good friend to my brother,' I say. 'He'll miss you sorely.'

'We will remember him in our prayers, you may be sure. We'll follow his work with interest and fully expect that his efforts will glorify God. And you also, as his devoted sister. It is a consolation to those of us who are his friends that you are going with him.'

It's nice to think that our friends will be keeping a vigil over our safety, I think. I need to put more faith in the power of the Lord, just as John does and our mother did.

'And further good news, Mary Anna,' John says. 'Mr Waterhouse has been appointed as General Superintendent of the South Pacific. He and his family will accompany us out there, and I'll report directly to him. He's guided me this far, and I'll be very glad to be part of his mission.'

So, it is decided, I think. 'Are you not tempted to join us on this adventure, Mr Barrett?' I ask.

'A little part of me marvels at what such a calling might be like, but my body does not lend itself well to extremes.'

I can see that might be true. He's a cerebral sort, without practical skills. When we walked to chapel earlier, I saw him picking his way along the cobbles, skirting the horse mess fastidiously. John and I were brought up in the country, experiencing nature and dirty hands, but Mr Barrett has probably lived in a city all his life.

'Well, we shall be glad to have a friend like you at home to send us news,' I say.

'Indeed,' John says. 'We give thanks to the Lord that His plan is coming to fruition.'

If only I could be sure that it is His plan, I think, and not just John's.

It seems that New Zealand is becoming a popular topic of discussion. In May, while my little black bee friends are busy with the bluebells, the *Times* newspaper likens the rivalry between the Wakefields and a certain Baron de Thierry as being like the feuds of York and Lancaster. '*Like bloody brothers fighting for a birth right*,' is how they put it. I hope it doesn't result in a similar war, especially if I'm going to be living in the disputed territory.

'Who is this Mr Wakefield I keep hearing about?' I ask John at dinner.

'Edward Gibbon Wakefield, who is not long out of prison. He's the ringleader, along with his brothers. I've read that he describes the natives as a barbarous people who can scarcely till the earth. Our Mr Beecham has joined forces with the Church Missionary Society to protest at his analysis. I don't believe Mr Wakefield has even been there.'

'The CMS? They're the Anglican missionaries, aren't they?'

'Yes, that's right. They're quite established in missions on the eastern side of the country, while our Wesleyan missions are to the West. It is an amicable arrangement.'

'And Baron de Thierry? What does he have to do with New Zealand and the Wakefields?'

'He claims that he bought land from some native New Zealand chiefs when they visited him in Cambridge a number of years ago, and that they wanted him to be Sovereign Chief. He's tried to get the British, Dutch, and French governments to validate his claims, in exchange for being appointed Governor, but to no avail. More recently he's been in Sydney, raising an army of colonists.'

'Why are the missionary societies getting involved, though? We Wesleyans and the CMS.'

'We're concerned that the rights of the natives are being neglected. Mr Beecham, as one of the secretaries of the Wesleyan Missionary Society, has published a book looking at the proposals for colonizing New Zealand. And there's another book advertised in the paper, on the topic of living in New Zealand aimed at prospective settlers. I think we've chosen a good time to go, Mary Anna.'

I'm not so sure. It's a different proposition to building up the congregation here in Birmingham. John has recently had some difficulties with a few of the young men he's been instructing, but at least they can understand his sermons and arguments.

A couple of weeks later, John is reading in the newspaper about the progress of the New Zealand Colony Bill, which has been debated in Parliament but thrown out on its second reading.

'I agree entirely with that analysis,' he says, folding up the paper and tossing it to one side. 'The editor was applauding the fact that the bill didn't pass.'

'I still don't understand,' I say. 'Aren't they wanting to do the same thing we're planning? Travel to New Zealand and establish themselves there?'

'Not at all, Mary Anna. We are interested in the souls of the natives there. These people, even though they number quite a few members of Parliament, are interested only in themselves. They hope to make money by buying land cheaply and selling it on at greater expense.'

'But surely it must be a civilising influence if more Englishmen are living there?'

'You would think so, and there are some grand names associated with the company, such as John Lambton, who's recently become Lord Durham, Sir William Molesworth, and William Hutt. But if the Bill had passed, it would have given the Wakefields and their allies unlimited power to make their own laws and rules, which I think would have been a dangerous move.'

It doesn't give me any comfort to know that even in Parliament there are differences of opinion on how to proceed. I'm heading to a lawless country, with powerful people focusing their attention there. I hope our own allies in the Society will continue to look after our interests.

'The Society has found us a ship,' John tells me, reading a letter that's just been delivered. 'It's called the *James*, and we'll be sailing in September.'

'So soon? That's only two months away. How will we ever get ready and make our farewells by then? Which port do we leave from?'

'Gravesend, on the Thames,' John says. 'So, you'll get to see London as we pass through. We must ensure that we have some time to visit the attractions before we depart.'

We pack our belongings into trunks and boxes. I deal with the practicalities of things like soap and linen, cooking pots and material. Cook spends more time with me, talking about how to keep weevils from my flour, moths from our clothes, fleas and lice from our bedding. I pack a box with seeds for a garden, hoping the mice and rats on the ship won't find them. John is concerned chiefly with which books and papers to take. Our supplies pile up.

'I think we should take Nancy,' John says one afternoon.

'The maid?' I ask. 'Would she be happy to come with us?'

'Why wouldn't she?' John says. 'You're prepared to go with me. Why would it be any different? Besides, we need someone to do the washing and cleaning, both on the ship and at the Mission House.'

'I thought they would have native girls to do that,' I say. 'Do you want to ask her, or shall I?'

'You're the housekeeper,' John says. 'I hardly ever speak to her. I think you should.'

I spend the night fretting over how to talk to Nancy. In the end, it's relatively easy. I approach her as she is resetting the fire in the drawing room. She settles back on her heels and looks up at me while I outline our suggestion.

'Cor, yes please, Miss,' she says, rising to stand before me. 'I was about to look for a new position, for I don't want to return home when you finish me up. I've got to liking me own bed, not sharing with me sisters.' She twists her apron in her hands.

'It may not be easy,' I warn. 'It will certainly be different to here.'

'Nothing's easy though, is it Miss?' With a grin, she returns to her task.

Indeed, there are various difficulties in preparing to leave. I spend hours trying to decide what I should take, and what I should do with the things I'll leave behind. Will I regret giving away that candle holder? If we return in a few years, which items will we want to reclaim? I carefully pack Jane's silver hand mirror which our parents gave her for her sixteenth birthday, and my mother's mourning brooch, mementos I'm determined to keep close by.

John is looking surprisingly morose when I come upon him at his desk one day. He's been energetic and enthusiastic as we make our way around our friends, but now he looks shaky.

'Are you certain about going?' I ask. 'I fear this preparation is taking too much from you.'

'I feel as if I could retire to a corner and weep, if it wasn't for your cheerful company,' John says. 'I have a knot in my chest, right by my heart, at parting from my old, tried and dear friends.'

John has always enjoyed the company of his fellow ministers. I have my own friends among the congregation, but no-one who's ever come close to replacing my sister Jane. That's a hurt that forever aches in the back of my mind.

'I'm sure it's not too late to change your mind,' I say. 'You could explain to the Society that your health has never been strong.' I feel a bubble of hope rise inside me at the thought he might undo his decision.

'No, I don't regret my task. We'll be doing the most honourable work; spreading scriptural Christianity throughout the world. What is my feeble body compared to that?'

The bubble bursts. He doesn't ask whether I'm sure I want to accompany him. He's focused on the work ahead, work which he's confident will lead to glorious reward. I don't know what I'd say if he questioned me thoroughly. Leaving isn't something I want to do, yet I feel I don't have a choice. I save my tears for my lonely bed at night and smile determinedly during the day.

Our house in Birmingham is packed up and we've farewelled Cook and Nancy, giving the girl strict instructions on how to meet us in London. Then we travel to Thirsk to say our goodbyes to our father and his new wife, Elizabeth. It feels very strange, staying in our childhood home with someone else in charge. Mother's things are still around, but Elizabeth has moved them about. It is disconcerting. Also, I'm not sure how to address Elizabeth. Calling her Mrs Bumby feels odd, saying Elizabeth is too intimate, and she isn't Mother. At least Father seems happy with the new arrangement, and that's what matters, after all. He'll need someone with him, since both his children are about to desert him.

Father is nearing sixty, quite an old man, and might not be alive by the time we return. If indeed we ever do. The thought that I might never see him induces a type of mourning in advance. My heart is sore, as if it is bleeding. The feeling brings memories of my mother's passing, and my sister's illness and death, and the losses I felt then.

'There, there lass,' he comforts me as my tears fall. 'I'll write ye often and be comforted by the thought that you and your brother are together.' His muscular arms from a lifetime of butchering are warm about me.

'I wish I'd been born a second son for you, so that I could remain and take over some of your work,' I say. 'I haven't been much use to you, have I?'

'Nay, ye have been a delight to me, Mary, with your kind heart. I would never of wanted otherwise.'

Thirsk is still the same market town as when I was a child, with its large, cobbled square and ancient market cross. Coming back after several years away makes it look even more peaceful and clean. Such a contrast to Birmingham, which has a hundred times more people, and where the soot from the constantly clanging foundries hangs in the air and covers every surface. Here, I can breathe, and I still know almost everyone. It feels safe and unchanging.

I take a stroll with John, my arm tucked into his. We don't speak as we wander past the so-familiar Wesleyan church, over the bridge which crosses Cod Beck, hurry past the usual carousers at the Black Lion, and up to the Town Hall. I am trying to lock my happy childhood into my memory so that I can recall it whenever I feel low.

Jane and I picked flowers from that field to weave into a daisy chain, pretending to be princesses. There is Thomas Lord's house, the famous Thomas Lord who went to London and built a cricket ground. I remember how the town mourned him when he died a few years back, even though he never returned here to live. A famous son, nevertheless.

Over there are Mr Webster's hives, hiding in their protective boles in the brick wall which shades the skeps from the worst of the wind and rain. My mother accompanied me on the outings when Mr Webster taught me how to calm and gentle the bees, and how to heft the skeps to check their weight. When Mr Webster died, I came back here to whisper the news to the bees and cover the skeps with black cloth. I wonder whether his son is still looking after them. Young Mr Webster didn't take as much interest in the hives as I did. The bees should be bringing in lots of nectar from the lavender and cornflowers I can see flowering everywhere.

On our way back to the house, we stop at the graveyard to visit our lost family. John puts his hand upon our mother's headstone and closes his eyes, as if he is communing directly with her. Perhaps he's praying. I leave him to it, and instead sink to the other side of Jane's plot and start picking out weeds. It's hard to believe that eight years have passed. How different my life might be now if my childhood companion had lived. The confidences we would have shared. Jane had ever been the prettier sister, and she might have married, with me the helpful spinster aunt caring for Jane's many children.

Jane's headstone includes a tribute to our oldest sister, also named Mary. She died at three, just after John was born. No wonder our mother cherished John so dearly.

Since Jane's death, I've planned to be buried with both of my sisters when my time comes. In the early days after she was gone, I sometimes imagined it was me in the grave with her, not our baby sister. Until she got ill, we shared a bed. It was lonely sleeping without her. Even now, years later, it still doesn't seem fair that such a sweet and vibrant girl died and left me behind.

I take a deep breath, raise my face to the sun, and say my goodbyes. Still, I am grateful for my faith, which tells me that I will meet these beloved ones again in that happy place above. While my mother's strong faith was firmly passed to John, it does also help to sustain me. How fortunate I am to have the prospect of eternal life beckoning me onwards. I'll have to keep that in mind in the times ahead.

That night, as I approach my old bedroom for perhaps the final time, I'm taken back in time to when Jane was alive.

I'm standing outside our bedroom door, listening to her cough. It's so bad that it seems as if she will never be able to draw another breath. I crave to go in and comfort my sister, as I did when she first became sick. I want to make sure her pillows are elevated, that she is warm and dry, not feverish and sweating. I remember brushing her beautiful black hair, before it became so lank.

This used to be our bedroom. Now I occupy John's room while he's away at the academy in Leeds. Perhaps it's a blessing that he isn't here to witness her contagion. He's always been inclined to chest trouble himself.

Mother rushes out of the room, and I catch her sleeve as she passes. Ma looks startled. I see she's carrying a bowl of dark red liquid – the blood Jane has been expectorating.

'Oh no, surely not!' I cry. 'It can't have advanced this quickly.'

'Now you understand why you cannot be with her,' Mother says. '*I will not lose both my daughters.*'

She hasn't closed the door completely. I glance through the crack and see Jane. She looks nothing like my younger sister now. She's fashionably pale and her dark hair emphasises that. Her arm, lying limp on the blankets, looks to be bone draped in skin with no muscle left. I'm shocked at the change from two months ago. All the strengthening broths that I've been preparing with Cook haven't made any difference.

This shouldn't be happening. We're a godly family, attending Church regularly, reading the Bible, living by His Word. Our brother is studying to be a minister. We live cleanly and wholesomely, not indulging in alcohol or tobacco. I thought the Lord would look after us.

Maybe it's my fault. Perhaps I haven't prayed sufficiently hard or believed strongly enough.

Last summer, when the cholera arrived, our family prayed fiercely, singing our hymns joyfully to show our faith to the Lord. Thousands of afflicted died in the big ungodly cities, but God spared us all in our little town, even the idolatrous Catholics. Jane was by my side then, soprano to my alto, thin but still with some health in her. We'd held hands, squeezing each other's fingers to show our mutual support. Our faith worked then. Why doesn't it now, when we need it more?

Mother scurries off, perhaps to consult with Cook or Father. She's not here guarding Jane, like she usually is. I know I'm forbidden to enter, but something rises in me, urging me to disobey my mother. It's temptation such as I've never experienced before. Maybe the hand of Satan. I'm not strong enough to fight it.

I tiptoe into the room, trying not to disturb her. I stand there, noting the blue veins in my sister's face and her paper-thin eyelids. Then the lids flutter open and I'm caught staring. Discovered disobeying.

'Mary,' Jane whispers. 'How I have longed to see you.'

I snatch up her hand. It's gossamer light, like the spiders' webs we used to collect to bind Father's cuts. I have no regrets now about entering the sickroom. She needs me, for my strength and support. Mother has been wrong to ban me. What does it matter if I succumb to the consumption if I can help my poor sister?

Mother's very angry when she finds me, but her face softens as she watches Jane sleeping peacefully.

'I do admit that she seems better with you at her side,' she says. 'And it's a relief to have someone else to sit with her.'

The next few days show an improvement. She has more vigour, chats away with Mother and me, and even speaks about leaving her bed. Her eyes sparkle with life, and she finds our tales of the outside world amusing again.

'It's so nice to hear Jane laughing,' Mother says. 'I wish I'd let you help earlier. It feels as if she's returned to us. I praise the Lord for His blessing.'

Our elation is short-lived, though. The following day, she is even sicker. Every breath is an effort, and she's unable to speak. She weakens as the day progresses, and by nightfall we know she won't last. She's so tired. Her breathing becomes slower and shallower. Father joins us at her bedside. There's nothing we can do for her.

I try to sing her favourite hymn: *Rejoice, The Lord is King*. Mother joins in, and then Father. It helps to fill the silence between her breaths. We see her mouthing the words. Then a small smile hovers at her lips. She takes a deep breath without coughing, and I wonder whether we've effected a miracle by invoking the Lord. However, that is her last breath. We sing on, watching with disbelief as her chest fails to rise and her face relaxes.

Tears stream down Mother's face, and Father sobs mid-sentence. I stop singing and begin to moan. It comes from inside me, unbidden and unwelcome, but unstoppable. I've failed. If only I'd been allowed to care for her throughout her illness, I'm sure I'd have been able to keep her alive. Now she is gone. I have no sister anymore.

It's no use blaming God. I don't need Mother to tell me that. He has His own reasons for taking her Home. No, the fault is my own. I did not sufficiently devote myself to Jane's care, and now I'll have to suffer the consequences. My selfishness had been punished, and my parents are being hurt as well.

The only way I can atone for my sins is to devote myself more fully to my remaining family. To my grieving mother and father. To my only remaining sibling, my saintly brother John. I'll serve them constantly, selflessly, and that will keep them safe. I neglected Jane in her need, but I won't make the same mistake again.

I lift her limp hands onto her chest, one over the other, and make her a solemn promise. 'I am the remaining daughter, and my family will know they can depend on me.'

CHAPTER 3

My hives are laden with honey. The bees have passed a bountiful summer collecting their supplies, and I've purposely decided to leave all the sticky sweetness to them, providing sustenance for my girls on their journey. The load of honey makes the skeps heavy, and I need assistance to lower them into the half hogshead barrels which will protect them on their journey. Oilcloth coverings shield them from inclement weather and help to keep errant bees enclosed. I've also put muslin cloth over the open bottom end of each skep to contain them while allowing some airflow, and they sit on a bed of straw for cushioning. They've just been loaded into the cart, along with most of the trunks and our furniture.

I hope that the jolting of the cart doesn't break any of the comb. I've spent a few hours hammering in more hazel spleets, sticks which are meant to stabilise the wax comb which hangs in great sheets from the top of the skep. The bees protested at me shaking them around, but my veil and gloves protected me. Now I'm concerned that the heat of summer will soften the comb, causing it to break. Or that the bees might overheat to the point where they die.

'You'll need to go slowly and smoothly, without too many bumps,' I tell the carter. 'And if you could rest the horse and cart in the shade from time to time, that would be good.'

He wipes his sweaty brow with his sleeve. 'Well now, I's not sure about this at all,' he says. 'What if them bees gets out? They'll sting the blazes out of me. And me horse might bolt. Then where would we be?'

'I don't expect they will,' I say. 'I've covered them up well. But that's why you should take it quietly.'

'It'll make the journey longer,' he says. 'I 'spect it should be worth more.'

'Listen, my man, we've already agreed a price,' John interjects. 'It's a good long journey, and you're to be well paid for it. I suggest you stop your grumbling and get along. We'll see you in Gravesend in two weeks.'

We are going by coach to Liverpool first, where we plan to visit friends. I wish I was accompanying my bees to keep an eye on them, but I'm also pleased to be travelling in a well-sprung coach with padded seats rather than a rattly cart.

The skeps will take up space and plenty of my attention on the long journey to New Zealand, so I've restricted myself to only taking two hives. In the past, I've found that having at least two colonies gives me more options. If one falters, then I can provide it with comb or bees from the stronger hive.

I've never migrated my bees onto heather or other blooms on the moors, like some beekeepers. I've been happy to leave the skeps in one place and allow the bees to forage close by. Now I wish I had more experience with moving them. With their cloamed hard outer shell made from plastering on cow dung, the skeps appear robust and waterproof, but I haven't properly tested them. I hope they'll survive the long journey ahead. It would be terrible to go to all this effort and have the hives fail. I am responsible for their lives.

I watch as the carter sets off, his shoulders up around his ears as he shrinks from his load. I'm sure he'll get used to the bees' presence during the journey. Bees rarely sting, in my experience, unless they are threatened. And I've chosen the quietest hives, with their dear little black bee occupants.

It is early September, the days shortening and becoming cooler, when we rattle into London. The noise of the wheels on the cobblestones and the jarring leaves us in no doubt that we are in a big city. I look out from the stagecoach windows, astonished at the size of the building opposite where we are pulling up. The white marble columns stretch up almost as far as I can see.

'The General Post Office,' the man sitting across from me says, noticing my interest. 'They built it ten years ago, and they reckon it'll be a White Elephant.'

'White for the stonework?' I ask. 'But why an elephant? Because of its size?'

'Too large for its purpose,' he explains. 'How many people will continue to write letters? Especially with these railroads opening up everywhere. We'll be able to visit, instead of corresponding.'

I nod, not keen to be drawn into yet another of the man's conversations. After twenty-two hours trapped with strangers in the coach, having only infrequent stops for quick meals while we change the horses, I'm ready to disembark and rest.

The following day we are still feeling worn out after our journey around the countryside and having to make so many sad farewells. However, we force ourselves to entertain our friends William and Sophia Hyde, who've come to spend time with us. Altogether, we set out to explore the great city. Our first stop is to visit St Paul's Cathedral, which is visible from many parts of the city.

'I know it is an Anglican church,' John says, 'but it was built before our great Mr Wesley showed us the better way, so I submit that it's perfectly fine to admire this monument to our Lord.'

'It is glorious,' admits William, 'but for my part, I'm interested in seeing the Parliament buildings. The papers report on their activities so often, I'd like to picture where they're placed.'

'In fact, we must walk down Fleet Street to get to Westminster, I believe,' John says. 'Past the source of our newspapers. Are you two ladies happy to continue to walk, or should I hire a hackney coach?'

'I'm glad to be walking, rather than sitting, after our long coach journey,' I say. 'You know, I thought Birmingham was a grand city, with all its industry, but London is so vast and incredibly crowded and busy.' I shrink closer to my brother as a group of rowdy young boys passes us. 'It reminds me of my hives on a summer's day. So much bustle, with everyone seeming to have a purpose.'

'It's also smelly and dirty,' Sophia says. 'I was looking forward to seeing the famous Thames, but instead it's muddy and rotten, more like a neglected pigpen. I don't know how those children can stand to be in there, picking through the debris.'

'I've read about that!' I say. 'One of those stories by Boz. Was it Oliver Twist? You know, *'The Parish Boy's Progress'*. I've been reading every instalment in *Bentley's Miscellany*. That's one of the things I'll miss most. I've just discovered that Monks is plotting against Oliver, and now I won't know what happens.'

'I'll collect those magazines and send them to you,' Sophia promises.

'Would you?' I ask. 'I'll so look forward to that.' Every month I've eagerly awaited the next episode, and I haven't liked to think that I might never know how it ends. John doesn't like the tales, saying they are frivolous entertainment and not true. But even though they are just made-up stories, the author makes it seem as though the characters are real people.

'There is a very yellow look to the light, isn't there?' William says to John. 'I understand that in winter it can be difficult to see your way, and wanderers find themselves in the river by mistake.'

'I don't believe Man was meant to live in these conditions,' says John. 'Millions of people crowded into such a small area. No wonder there is a surfeit of contagion here. All those people who died from cholera. I look forward to the uncluttered environs of our destination.'

'When does your captain expect to leave?' asks Sophia. 'I was so hoping to wave you goodbye, but I fear we must return home within the next few days.'

'He says it will be another week before the tides are suitable,' John says. 'I am to give a valedictory service in the Queen-street chapel here in London before we leave. I hope you'll be able to attend that.'

'We would be honoured to hear you one last time,' William says. 'But after that we must return to Birmingham, unfortunately. Still, you know you go with our warmest wishes. Come now, my dear, don't fret. The time to part is not yet at hand.' Sophia seems close to tears, and William puts a comforting hand on her back. 'Look, here is a street with no gas lights. Birmingham is not so far behind London after all, since we have street lighting in almost all places.'

After a tearful farewell from the Hydes, we take a steamer, the *Gem*, down the river to Gravesend, accompanied by Nancy. The paddles of the boat splash through the water while soot and steam rise from its chimney. I watch as both sides of the riverbank slide past, their banks filled with warehouses and fields and grand buildings. We pass

Greenwich and John points out the ball suspended on top of the octagonal brick building, set to drop at one in the afternoon so that captains can set their chronometers. Our steamer chugs under several beautiful new bridges that span the river.

'You know they're building a tunnel under the river?' John asks me.

'No. Really?'

'They've had several tries at it, I understand from the newspapers. Somewhere underneath us there may be Cornishmen digging away.'

There seem plenty enough bridges crossing the Thames, without the need for a tunnel. Still, I reflect, there are a lot of people in London. John says two million, more than ten times the number of residents in Birmingham.

At Gravesend I am finally able to see *James*, the vessel where we'll probably spend the next five months imprisoned, as I think of it.

'It's a barque,' John says.

'How can you tell?' I ask. Neither of us has ever been to sea before.

'Look, sister. It has three masts, two with square-rigged sails.'

He is full of knowledge, and I don't understand how he has gleaned it.

The *James* looks rather small compared to some of the other ships scattered around the water. It's hard to see how it can survive the seas or how everyone is going to fit aboard comfortably.

Captain Todd is a rough-looking man with a great bushy beard, and I'm relieved to hear he is a Wesleyan and apparently very nice. There are twenty-one cabin passengers in all, mostly fellow missionaries, including the Waterhouse family. Nancy, of course, will travel in steerage, and I descend to inspect that area as well. It is crowded and dark, not at all appealing. Nancy must be grateful that she will spend much of her time in our rooms.

Most of my supplies are already aboard and stowed, and only my immediate needs are kept at the inn where we are to stay before departure.

On our last day on shore, our friend Miss Emma Sells comes to visit. Rain keeps sweeping through the area, so we confine ourselves to the sitting room of the little inn from where we can see *James* at anchor.

'Dear Miss Bumby,' Emma says, taking my hands. 'And Mr Bumby,' glancing at John. 'The last occasion when I was in London was such a joyous event, and now I am full of sorrow in seeing you leave.'

'Oh, yes,' I say, not wanting to dwell on our imminent departure. 'I'd forgotten that you came to London for the Queen's coronation. You must tell me all about it.'

'It truly was a great spectacle,' Emma begins, sitting back and smoothing her skirts. 'Father found us an acquaintance with a house in Grosvenor Place and I was able to sit on a balcony to see the procession. Everyone was in their best attire and the whole city was in such a mood of joyful anticipation and good humour. We spent our time figuring out who all the great dignitaries were, and cheering as they passed, and then when Her Majesty's carriage appeared, we ladies were so excited we took off our bonnets and waved them about.'

Emma continues with a description of what she saw of Her Majesty's clothing and entourage, and what she herself had been wearing. My thoughts drift off. It seems impossible to believe that I'm leaving the greatest city in the world, perhaps forever. My imaginings of the place we're headed to are indistinct, but I don't doubt that it is going to be primitive.

Eventually Emma discovers that she's lost her audience, and John kindly suggests we take advantage of a fine spell of weather to walk about the town. However, we find it dirty and dishevelled and are glad to return to our accommodations, where we bid Emma a final farewell.

I find that I have no tears left. To soothe my nerves, I constantly remind myself of the purity of my actions. The thought of stormy seas and the peril we are putting ourselves in makes me quake, but I resolve to place my trust in our Lord. He will not fail us, not when we are on His mission.

The next morning, I am astonished and delighted to see Sophia Hyde arrive back.

'We reached home but I felt so unhappy about having failed to see you sail,' she says, 'Mr Hyde told me to take my maid and come straight back. So, I have.'

We take a small craft out to *James* and about noon the rest of the party and friends come aboard for a farewell dinner. Copious quantities of roast beef, curried chicken, and fish fillets, along with plum pudding, are laid before us. I don't have much of an appetite, the continued stench of the river mud being off-putting, but the festive atmosphere is contagious.

After the meal a service is held on deck, taken by the senior Missionary Secretary, the Reverend Jabez Bunting. It is as if I am in the presence of Wesleyan royalty. The chorus of our voices sings the hymn, '*Hail the Day That Sees Him Rise*'. Our unity of purpose shines through the men's deep bass and baritones, but I can hardly hear my own alto tones. The song both steadies my commitment and causes further afflictions to my heart. I feel as if the moment isn't real; that this is happening at a distance, to someone other than me.

Sophia tugs on my arm. 'My dearest Mary, return home with me,' she says quietly. 'I have such a bad feeling about this. I don't think it is right that you should go.'

'Oh, I wish that I could stay,' I say, the tears springing to my eyes again. 'But it is too late, and I am intent on doing my part.' How dreadful it would be to change my mind at this last moment, in front of all these brave adventurers. How could I ever explain it to John, even though I'm tempted?

Then Sophia and the other passengers' friends retire to the shore, waving and calling out their goodbyes.

'Adieu, dear one,' I cry to Sophia amongst the noise. 'Remember me.'

'Fare well,' I hear the reply across the water, 'Dearest of friends.'

The ship weighs anchor, the sails start to unfurl, and we set off down the Thames.

CHAPTER 4

As the pilot vessel guides *James* down the Thames, I can't bear to leave the deck, thinking that this might be my last glimpse of home. The water is calm with an oily sheen, the slight breeze riffling it in places. Along the banks there are still scattered groups of the poor, especially the young and very old, scavenging detritus from the river's edge. Large white birds with spectacular wingspans wheel above us. Their squawks add to the cacophony of the many sails flapping and the sailors' calls.

The other passengers appear to feel as melancholy as me. There is such a crowd on the deck watching the land slip past that the sailors are finding it difficult to move about.

'Take care!' someone shouts, as an oblong wooden object falls from the rigging.

I am jostled as people try to move. The large missile hits two of the sailors, who drop to the deck injured. There's a confused period as the sailors are examined and then carried bleeding from the deck to recover. I take the chance to retreat to my cabin to mull over my thoughts.

I haven't been on a ship before, barring the steamer, and it is a strange place, unlike anywhere I've ever stayed. It is cramped and crowded, with narrow passageways and steep stairways. The timber feels damp and sticky, and doors need an extra shove to open. The names of the decks and various areas are like a foreign language. Someone mentions that the projectile which fell was a block, but that doesn't mean anything to me, and I'm not sure whether that sort of accident is something to be expected in the future.

My hives have been placed on the poop deck, secured by rope to prevent them from tumbling about. There are cages of fowl there also, chickens and ducks, while larger animals are somewhere below deck. I can hear their unhappy bellowing and bleating. The captain has mentioned to me that some of the sailors are unhappy at having bees aboard, concerned that the insects might escape and start stinging. I assure him that the bees are safely enclosed in the hogsheads, and that I've stuffed corks in their flying-opening so that I can control when the bees exit. The usual dried grass or moss won't last the distance.

When the hives were delivered to the ship, I'd made additional holes in the skeps to help the bees gain sufficient air and so that, later, I can drizzle sugar syrup in. I'm still worried that the bees will overheat, or starve, or chill. Amongst my supplies I've added extra sugar so I can feed them during the journey. Whenever I visit them on the poop deck, I'll loosen the bindings and heft the barrels to accustom myself as to how heavy the hives are. I hope to learn when they became light and require that sugar.

The ship is still in sight of land in the evening when we missionaries gather on deck again for another service of worship. It feels as if we are returning to the pattern of John Wesley himself, who did so much of his preaching in the open air. It is pleasant to leave the stench and congestion of London; to be moving down the river and into a fresh breeze.

I retire for my first night aboard, but don't sleep well. I'm disturbed by the strange motion of being at sea and the various creaks and clangs of a ship under sail. There's a lapping sound that makes me worry that the ship is taking on water. My thoughts buzz around the fact that I'm leaving my friends, my father, and all I have ever known.

When I arise, I find that light winds overnight mean the ship has not made much progress. The pilot is still on board, guiding us through the passage of the Thames. After breakfast, I spend some time in my cabin writing further letters of farewell. I entrust them to the pilot when he leaves to return to Gravesend. I don't have any fresh news to give my friends, of course, but find it useful to put my thoughts to paper.

As we make our way, I find I'm gaining confidence with the roll and sway of the ship. As I walk along a passageway, I no longer feel the need to press my hands on the walls. In more open areas, I don't cling to furniture to keep steady. I feel that if our voyage continues like this, it will be tedious but acceptable.

Two days later, my view of the voyage changes. The ship encounters extremely rough weather and I take to my bed. I'm not alone in feeling indisposed – all the passengers are similarly affected, including John, but he's well enough to visit me.

'Where's Nancy?' he asks, after he knocks and enters my cabin.

I look up at him through bleary eyes. 'I imagine she is in the same position,' I say. 'I haven't seen her since I came abed.'

'I thought her place would be with you, though,' John says.

'When will this end?' I ask, as another wave tips us around. I'm embarrassed to be cradling my chamber pot, which is slopping with the late contents of my stomach.

'Captain Todd says that it is unusually rough, so he plans to call into the Isle of Wight for some respite,' John says. His face is pale and sweaty.

'I don't know if I can take much more of this,' I say, as another heave sends me hovering over the pot.

'The sailors talk about getting our sea legs,' John says. 'The sickness usually disappears after the first week, as we become accustomed to the motion.'

'A week! If only I could sleep that time away and wake up well again.'

'Here, sit up and have some water,' John says. 'I will take your pot to be emptied. The very smell of it must be making you worse.' As he lifts the pot away, he pukes into it himself. He braces himself against the wall as the ship pitches, and staggers into the passageway with the handle clutched in one hand.

I can't tell him that I have soiled myself as well, during one of the worst episodes of retching. There has been nothing I can do about it, and I am still so miserable I don't think I can clean myself. He can probably smell it but is too polite to mention anything.

The thought of reaching the calm and safe harbour of the Isle of Wight keeps me hopeful. Then Nancy might be able to assist me.

When the ship eventually reaches Cowes, a dozen of the men venture ashore for breakfast and to purchase some small items they've forgotten. They come back with news of the island.

'They say that the new Queen comes here for holidays,' one says. He indicates a castle on the hill, just visible through the masts of the numerous other ships in harbour. 'Maybe she stays up there,' he suggests.

'She must have a stronger constitution than me!' says another, pulling the leg off a crab he's purchased, and drawing out the tender flesh with his teeth. 'That passage near did for me.'

The sight of the men eating makes me feel queasy again, despite the calmness of the harbour. I take another sip of water, still unable to countenance any food.

I'm not the only one feeling so bleak. One of the gentlemen aboard decides that he can't face the complete journey and announces that he'll leave the ship and find another passage home. I watch enviously as all the passengers assemble on deck and give him a rousing cheer.

The great noise we create startles a pig which has got loose, and it races about the deck and then tumbles overboard. I hold my stomach, aching from the previous day's retching and now sore from laughing, as the sailors rush to rescue the animal. Then there's a sudden torrent of rain, and we chuckling onlookers scramble for shelter. It makes a pleasant change from the worry and concern which preceded this journey. Maybe the worst is over.

The entire contingent turns out to see Lizard Point at the end of Cornwall, which is truly our final view of our homeland. It is such an unsettling feeling, moving away from all that is familiar and comforting. I never expected to do this, even in my wildest imaginings.

Then winds become more favourable and the seas turn calmer, although I am still unable to stomach much food. I spend a miserable birthday alone in my cabin. It's my twenty-eighth anniversary, but John has completely forgotten the significance of the day, and none of our friends are present to remind him. I wonder whether I'll survive this trip to celebrate my next.

Fruit is the only thing I can force myself to eat. Sweet and juicy, the end of the season's apples and pears keep me sustained. Unfortunately, over the next few weeks our fruit supplies diminish.

'Look, Miss Bumby, they tell me that island is Madeira,' one of the missionary wives, Sarah Ironside, says. 'Home of the famous wine.'

We watch as *James* sails within eight miles of Madeira. I can clearly see its white-painted houses.

'Oh, I wish we could call in there,' I say. 'My mouth is watering at the thought of the grapes that must grow there.'

'Are you still feeling poorly?' Mrs Ironside asks. 'I was fearfully unwell earlier on, but that has passed now.'

Her comment doesn't serve to console me.

My eyes continue to follow the island long after it becomes a dot on the horizon behind us. Then we pass the Canary Islands, slightly further from the ship's course so that details are harder to discern, and its delights are also unattainable.

It's strange to think that we are travelling down the western coast of the African continent. The land isn't visible at all, with the ship staying out to sea to catch the right winds. Each day as we move further from home the weather gets warmer.

'We should have stopped at Madeira, as you wished,' Mrs Ironside mentions one day, when it has been oppressively hot.

'Why do you say that?' I ask, as I take a sip of tea and wince at the taste.

'The wine they make there is specially formulated to last through this hot weather,' Mrs Ironside says. 'Not that I approve of alcoholic drinks, of course. But this water has turned most disagreeable.'

'Yes, it has a putrid smell,' I agree. 'I wouldn't like to drink it without boiling it first.'

When it rains a few days later, everyone scurries to leave basins and bottles out on deck, gathering the fresh water. It is a relief, but it doesn't last long.

Whenever I'm able, I join the Sabbath service on deck. It's heartening to hear the preachers speak about placing their trust in the Lord.

'Wasn't Brother Creed's sermon wonderful today?' John asks me.

'We're fortunate to have all these ministers on board with us,' I reply. 'Although I prefer your style of preaching.'

'You're just being partial,' he says, grinning.

'That may be, although it might also be true. You know, I can't quite get accustomed to worshipping under the sky, surrounded by all this unremitting sea.'

'I enjoy it. It makes me better appreciate the vastness of His power and authority,' John says. 'And we play such a small part in it.'

'That might be what I dislike. The feeling of smallness,' I say. 'Maybe I lack your trust in the Lord.'

Gradually, I become more able to participate in life aboard our small home. With my appetite returned, I join the rest for breakfast at eight, lunch at eleven, dinner at half-past two, and tea at six. Catering that number of meals keeps the cook and the steward busy. Counting all passengers and crew, we are fifty-five in number, so at the start of the trip supplies needed to be crammed in every crevice.

Although my illness abates, as we approach the Tropic of Cancer the heat becomes overly oppressive instead. At night I feel suffocated in my small sleeping cabin, with the evening's fog and damp. It's so uncomfortable that I lie there tossing in the sticky sheets, unable to rest. I wish I could take my bed on deck and sleep under the stars, like some of the men, except that wouldn't be proper. During the day I venture out, but I try to avoid being smitten by a burning sun.

'Nancy, am I cantankerous?' I ask one night, as the maid is folding away my clothes.

'I'm not certain what that means, Miss,' Nancy replies.

'Oh, peevish. You know, disagreeable,' I say.

'We all are like that these days, Miss,' Nancy says. 'What with the hot, and no sleep.'

I lie back into my pillow. 'That means I am, as well,' I say. 'How disappointing. I thought I would have more fortitude.'

'You do, Miss. Too much thinking, if you ask me. Was there anything else you needed?'

I fear for my bees and take every opportunity to release the oilcloth covers and open the hives. The other ladies aboard are horrified, and several of the sailors make protests, but it isn't often that anyone is stung. The bees don't go far beyond the boundaries of the ship and only perform short circular flights before returning to the skep. I keep the direct sun from the skeps by draping them with damp matting. I wish I could wear my bonnet similarly wet. The bees prefer to remain outside, clustering as a black beard on the side of the skep with their backs to the entrance and their wings fanning the air. When the ship is becalmed for over a week, I start leaving saucers of sugar syrup nearby for them. Gradually, the others grow used to the bees' presence, and some of the children become quite curious.

'Miss Bumby, why are the bees drinking water from the saucer?' young Joseph Waterhouse asks. He's about ten and frustrated that his energy has no outlet on board ship. 'We're surrounded by ocean. They could easily get water from that.'

'Dip your finger in the saucer and lick it. Do you see? It's sugar water. The sugar is their food because there are no flowers on the ship.'

'How do they get food from flowers?' he asks.

'Have you ever watched a bee visiting a flower?' I question him. 'Sometimes they rush around the top of the petals and get covered in a yellow dust. They take that pollen back to the hive and feed it to their babies.'

'They have babies?' Joseph asks.

'Of course, but they don't look like tiny bees. They're more like maggots.'

'How awful. Can I see?'

'No. They're protected in the hive. Anyway, sometimes a bee spends longer looking right into the heart of the flower. Then she is sipping up the sweet liquid in the base, called nectar.'

'Or he is. How can you tell boy bees from girl bees?' Then Joseph blushes. 'You don't have to tell me.'

I bite my lip to stop myself from laughing. What city folk these Waterhouse children are. 'All the bees who are coming out to find food are girls,' I say. 'You can tell the boys because they are larger with huge eyes.'

'Like men are larger, and women do all the work around the house,' Joseph says, and nods. It fits well with his view of the world.

'Mhm,' I agree. I don't say I think male bees are almost useless. There's only so far you can draw parallels with people.

'Oh, this weather is so hot,' Mrs Waterhouse says later that day. She's always busy, with her large family. It's taken me some time to learn all their names. Two of the Waterhouse daughters, Mary and Jane, are adults, while young Mary Ann is only twelve and little Emma is six. The two youngest boys, Joseph and Samuel, are constantly under the sailors' feet, asking questions. The older sons are more self-contained, except for Jabez, who frustrates his father with his threats to throw himself overboard.

'Yes, even with my summer muslins on I feel as if I can't breathe,' says Mrs Warren, wife of one of the other missionaries. 'I wish the vessel would move, to create a breeze.'

'Actually, we need a breeze first before the ship can move,' says young Samuel as he passes. As one of the youngest aboard, he is indulged rather than corrected for speaking out of turn.

'Why don't we ask whether we can take a boat out for an excursion?' Mrs Waterhouse suggests.

'That's a lovely idea,' I say, and there are murmurs of assent from the other ladies. As Mrs Waterhouse is the most senior lady, she goes to organise it with Captain Todd.

I fetch my bonnet and a parasol. It will be too easy to end up with freckles or coloured skin from all the sunshine if I don't take care. The young boys have brown faces almost as dark as the sailors.

One of the boats is launched from the deck and we ladies are assisted into it. We spend nearly all day idling within hailing distance of the ship, attached by a rope. I find it very pleasant to be off the confines of the ship and closer to the water. I lean over the side and trail my fingers in the cool liquid, gloves safely in my lap, thinking of home. It's hard to imagine that winter will be setting in there now, with frosty mornings and bitter winds.

Overnight, to my horror, a large shark is caught by one of the sailors with a baited line, the first shark we have seen. I shudder to think I was dipping my fingers in the ocean where such a beast lurked. Cook prepares it for breakfast.

'How is it?' I ask John, as he takes a mouthful.

He chews for a short time, considering. 'Fine, nothing exceptional,' he says. 'Are you going to try some?'

I shake my head. 'I can't bring myself to eat such a monster which may have partaken of human flesh.'

'Oh yes, I can see your point,' he says, and pushes the uneaten portion of his meal aside.

We seem to be the only ones who have such concerns, as the rest of the passengers appear delighted with the fresh meat. I decide against sharing my reservations with anyone else so that they don't become uncomfortable with what they've done. I regret that I've put John off his meal. It seems that once the thought crosses your mind, it's hard to be rid of it.

The calm breaks, and I become queasy again. I'd make a very poor sailor. I can't understand how they stroll around the deck without a care. The ship is tossing and rolling in the rough seas, as if it is a cork upon a boiling pot, but at least it stays upright. A few days later we meet another ship which has lost one of its sails in a squall. It reminds me how fortunate we are to have such a sturdy vessel.

The weather remains hot; so hot that it melts the pitch from between the planks on deck. We can see a great number of sharks following and playing about the ship, their fins surfacing before they glide back out of view.

'I really don't like the sight of those fish,' I say to Eliza Creed as we watch the sharks swimming in the wake. 'It's as if the sharks are loitering, hoping for my demise.'

'Miss Bumby! You do have some strange fancies,' Mrs Creed replies. 'They are just dumb creatures. The Lord didn't make them with motives and schemes.'

'Still, they need to eat. I'm just glad to know that if I die at sea, the shroud will be weighed down so that it sinks out of reach of these vile creatures.'

'Pray the Lord that it doesn't come to that,' Mrs Creed says.

'Indeed,' I say. 'Even if I were insensate, I wouldn't want my body mauled by the beasts.'

Several days later a homeward bound ship from India stops to exchange information. It takes on our letters for delivery and its captain says he expects to arrive home by Christmas, which is in seven weeks' time.

'Christmas, Miss,' Nancy says, as she brushes my hair and ties it up. 'Our first one away from home. It does seem strange, don't it? And we don't even have no goose on the ship.'

'It will be a day of worship, nonetheless,' I say. 'We will be celebrating the birth of our Lord, after all.'

I notice that John is settling to the rhythm of the journey. He often exclaims about the brilliant sunsets we are experiencing as the ship nears Trinadad in the West Indies, where the captain hopes to take advantage of the trade winds to speed our voyage further south. John takes an interest in the management of the ship and ensures that we know our course, which I sometimes record. Meanwhile I am constantly laid low by the trials of life on the sea. I try to contain my complaints to my journal and not allow them to mar my letters or conversation. I feel as if this journey has reversed our personalities. Before, I was the cheerful sibling, while John was normally serious.

Wash day is even more of a chore than at home. It seems strange to be surrounded by water, yet unable to use much of it. Clothes washed in the salty water become stiff and uncomfortable, so we wear our outer garments for longer and use the precious fresh water supplies for our chemises. I help Nancy, since the other wives are without servants, and I don't want to appear lazy.

'Miss Bumby, do you think you could lock your bees away on wash day?' Mrs Waterhouse asks, holding up a white petticoat. It's no longer completely white but speckled with brown and gold flecks.

'I'm sorry, I didn't consider that,' I say. The bees have done a cleansing flight of their excrement over the clean washing, which was laid on the deck to dry.

'It's not a big problem,' Mrs Waterhouse says. 'It hasn't done any real damage. It just seems a shame to work so hard to get the clothes clean, only for them to get besmirched so soon.'

'I'll shut off their entrance on wash day,' I promise. 'I wish I could give you some honey in payment for the inconvenience, but the bees need it all.'

'Maybe you can send me a jar when the bees are established in New Zealand,' Mrs Waterhouse says. 'I'll be the most popular person in Hobart Town, with my special honey.'

'I wish you were coming with us to New Zealand,' I say. Mrs Waterhouse has been rather like a mother to me on this voyage, except that with her own large brood she doesn't have much time to spend with us other women.

'Maybe one time I will be able to accompany my husband when he visits,' Mrs Waterhouse says.

'Yes, I would like that,' I say. These friendships I've made on the ship are going to be torn from me. Mrs Waterhouse and Mrs Eggleston will both be leaving when they reach Hobart Town. Mrs Creed and Mrs Ironside will cross to New Zealand with us but will be posted to their own missions with their husbands. And as young women, they'll likely soon be busy with babies arriving.

It's going to be lonely, I realise. Only John and Nancy for company.

At least there will be plenty of work to keep me busy.

CHAPTER 5

We pass the equator.

It is an awful place. I can't imagine anyone enjoying the intense heat or the way clothes stick to the skin. The ship bobs on the water while the sails flap uselessly. No wonder getting through this area is so celebrated.

'It's a tradition amongst sailors that everyone is doused when crossing the line,' Captain Todd warns us. 'The sailors believe that it brings good luck from Neptune.'

'Does that include the ladies?' I ask. How can I preserve my modesty if I sport clinging wet clothing?

'Yes, ma'am,' the captain replies.

I wonder whether I should just remain in my cabin for the day, but it's so hot I can't bear the thought. Besides, I want to see what happens.

'Alright lads,' I overhear the captain telling the crew, 'I know you want your fun, but just remember that you have ladies on board. And even our gentlemen are restrained and cultured men, so you need to keep it light. I don't need any complaints.'

When I venture on deck, I see all the buckets and tubs lined up. When everyone is assembled, it turns into a good-natured battle, with water sprinkled over every sailor and passenger. Now I am pleased that the weather is hot so that my clothing dries easily.

I'm still unable to eat much, which is to my benefit when a few days later the cook serves saltfish and potatoes in what he calls a 'twice laid' manner. Either the fish is rotten or has been in close contact with some rat poison. Whatever the cause, twenty-three people become very ill. While I am exempt because I didn't eat the food, John is badly affected. He retreats to his cabin, complaining of stomach cramps. I visit him and am aghast to find that I can't wake him.

'John, stay with me,' I say, holding and rubbing his hand. 'Brother, don't forsake me now.'

I don't know what to do for him, and there are so many others who are ill that I am mostly alone with my thoughts. Nancy comes and goes with wet cloths, which we apply to a rash on his face and chest, and soups which we can't feed him.

'I'll stay with him, Miss,' Nancy says. 'You should have a rest.'

'I can't leave him, Nancy,' I say. 'He's all I have left. I don't think I can continue this horrible journey if he perishes. I just pray that the Lord won't take him from me.'

I feel very alone, out there on the great deep, far away from any medical assistance. I monitor his weak pulse and beseech it to continue. I listen to his shallow breathing over the creaks and groans of the ship. And I pray for his recovery.

Suddenly, I'm not sitting by his bedside. Instead, I'm back beside my mother, seven years ago.

'Mother,' I call to her. 'Mother, don't leave us.'

She struggles to open one eye, and stares at me. She seems bewildered. 'John,' she croaks through one side of her mouth. The other side droops unbecomingly.

I know she is calling for my brother. Although Pa is also named John, Mother always addresses him as Mr Bumby, or Husband.

'I have sent a message,' I tell her. 'He is in Halifax. But I know he will come as swiftly as he can when he knows you are unwell.'

I'm not sure whether she can hear me, or perhaps it is that she doesn't understand. Again, she moans, 'John'.

I don't know what to do to reassure her. Doctor Newman has been and diagnosed apoplexy. Half her body has been affected, and she can barely move. I try to give her a drink, but it spills from her crooked mouth. I end up spooning it in as if she was a young child. Doctor Newman has said to keep her calm and comfortable.

Pa comes in quietly. He kneels beside her bed and whispers prayers into her ear. I turn aside, awkward at seeing the emotion in his face.

Mother looks past Pa, at the door. I know she is looking for her only son. 'He will be on his way,' I tell her, although I'm not sure whether the messenger will have found him yet. I want to give her hope. 'Don't worry, Mother. I will look after John for you. I will care for him as you have done.'

Her eyes flick back to me, and though one is half-closed, I see that the other is brimming with tears. She has accepted my offer, my sacrifice. She stares a little longer, then closes her lids.

I stay beside her all night, watching and praying, listening for the sound of horse hooves on the stones or footsteps hurrying close. John hasn't made it home in time. In the early hours of the morning, she slips away from us.

I jerk awake in the chair beside John. He is still insensate, not stirring when I call his name or wipe his brow. I resume my prayers, hoping against hope that my brother won't follow our mother's path.

As night begins to fall, ten long hours after he was taken ill, he wakes and begins to vomit and purge. My relief is huge, and I'm overtaken by exhaustion. By the next day he's recovered, as have the others.

'I thought I was going to lose you,' I tell him. 'I couldn't wake you, and you were so pale and cold.' Tears leak from my eyes as I remember my terror.

'I'm sorry that it gave you such a fright,' John says. 'But it can't have been that bad. After all, every affected passenger has recovered.'

'It reminded me of Mother,' I say. 'Of how she didn't wake, no matter what we did. And then she died. I thought that was going to happen again.'

'Oh Sis, I would have woken to your sweet voice if I could,' John says. 'But the Lord didn't see fit to call me yet.'

'I'm sure my prayers contributed to your return to health,' I say. 'I just wish my own sickness would cure as quickly.'

We are now in November. I find it strange to have the heat and light of summer, when at home in England they will be experiencing short dark days. Furthermore, as the ship sails beyond the equator the days get longer, so that by the time we reach the Tropic of Capricorn we have daylight from five in the morning until seven at night.

We begin to look forward to reaching the Cape of Good Hope, where we can restock supplies. The last of the sheep has been killed for fresh meat. That upset the younger Waterhouse children, who'd made a pet of it. As a butcher's daughter, I am better prepared to accept the realities of slaughter. I know it is unwise to name and befriend an animal destined for the table.

After three months at sea, we finally see the top of Table Mountain, like a thin band on the horizon. All eyes are fixed on the hilltop as we sail closer, the mountain becoming clearer. It is riveting to see land again after so long with nothing on the horizon. The colours of the mountain, with its bare brown rock and the green of vegetation further down its slopes, are a relief after the constant blues and greys of sea and sky. As we get closer, we can also see the masts of many ships sheltering in the bay.

'Thirty ships, Miss Bumby,' shouts Joseph Waterhouse as he races past me. 'The whole of the South Seas must be here.'

I can hardly sleep that night for the thought of setting my feet on solid ground once again. The next morning the company is up and ready for breakfast at a much earlier time than usual, and by nine o'clock we have landed on shore. I am astonished to find that I can hardly stagger up the beach. My legs feel unsteady without the need to continually adjust my balance. I've dreamt of this moment, and yet it is most uncomfortable. Any time I close my eyes, the ground sways. My body seems to want to rock of its own accord.

On the beach we are met by Mr Hodgson, who has a mission house nearby with a chapel attached. 'Welcome, friends and brothers,' he says, shaking the men's hands. 'We've had newspapers from London, telling of your departure.'

'That means that another ship was here faster than us,' I whisper to Mrs Creed.

'Doesn't it feel strange to be walking on land?' Mrs Creed replies. 'And look at this town! It's huge.'

Mrs Hodgson is also very welcoming, and soon has a dinner ready for the crowd. Then we are shown around the sights of Cape Town. I am particularly interested to see grapes growing over the doorways of houses, but the one I try isn't yet ripe and I have to spit it into my glove.

The next day we travel by wagon to Constantia and see the famous wine-growing region. This time I get to taste the wine, as well as some delicious bread and cheese, before we finally arrive back in Cape Town.

'Thank you so much,' I say to Mrs Hodgson the next morning, as we prepare to leave. 'I only hope that New Zealand will be as welcoming and picturesque as the Cape Colony.'

It is strange being back on board the ship. It's both familiar and cramped, after the wide-open spaces of Cape Town. Still, I am eager to be on our way, so that I can finally stay in one place. I'm frustrated when a strong gale arises and stops us from leaving. However, it's just as well that we are still at anchor, because out of the dark we hear the cry, 'Lost, Lost.'

'Who is there?' one of the sailors calls out.

'Help us,' comes the reply.

A small boat is drifting against *James*. When our sailors get the people on board, they find five men and two women who had ventured ashore for supplies. This group tried to return to their ship, *Platina*, but the wind became too strong. The waves were boisterous, and their boat was being swept away. My fellow shipmates wrap them in blankets and keep them on board overnight until they can return to their own ship.

'Thank you, our dear new friends,' they shout, as they row back to *Platina*.

'They would have been swept out to sea and likely died, if not for our ship,' John remarks. 'Think where the wind was coming from. We are the last vessel in the Bay. Beyond us was oblivion.'

I shudder to think of being in a small boat such as theirs, floundering on the open ocean.

Several days later, we on *James* partake in another rescue. The captain of an American whaler, *Clematis*, hails us and comes on board. They've been twelve months at sea, and several of the men have scurvy. The captain's chronometer has broken, and he hopes to learn the latitude and longitude. Captain Todd gives him a basket of potatoes and pumpkins, and the passengers offer a large basket of fruit. Meanwhile, the sailors generously pass a few small comforts to the crew of the boat. Everyone on *James* feels that the circumstances could easily be reversed.

I count myself fortunate in my fate and thank the Lord for his assistance. In Cape Town we'd learned of one ship which lost thirty passengers after leaving England, and another which lost nine. For ourselves, earlier in the journey, one of the young apprentices fell overboard, but he was able to swim long enough for a boat to be launched to his rescue. Our worst day was during a tremendous storm, when one of the yards was broken and a sail split to pieces. The sailors took up all the sails and the ship went under bare poles. Other than that, we have not suffered greatly.

We spend another month at sea. I am very ill again, and waves crashing over the ship make my bed wet and uncomfortable. Christmas Day is a reminder of how far from home we are, and I wish fervently that I was back there with my friends. The year is coming to an end, and life is so different now to the beginning of 1838. The ship tosses on the waves, and my head aches even more when some of the gentlemen ring in the New Year by tolling a large bell. If only I could rise from my bed to ask them to stop. My diary slides off the desk onto the floor, and I must wait until Nancy visits to ask her to pick it out of the damp. It doesn't contain much of importance. I can never think what to write. Still, I should keep it safe, I suppose. But the very thought of leaning down to collect it makes me feel even more nauseous.

The noise from above is great, with sailors shouting and the chicken coops being thrown about by the motion of the ship. I wonder how my hives are faring. I hope the waterproof cloth that I tied over them is keeping the worst of the weather from the bees. How bemused they must be with this continual motion. It is difficult enough for a human, who can understand the situation. The two families of insects must be distraught by this change in their circumstances, with no explanation or hope of reversal. I wonder whether I have done the right thing by them, bringing them on such a journey.

Nancy comes to bring my tea, in the hope that I can swallow it.

'Miss, I'm sorry to tell you about Mr Bumby,' she says, as she helps me into a sitting position.

'No! What's the matter, Nancy?' I ask. I try to get out of bed, but I'm too weak to stand.

'He's hurt his back. Falled out of bed, he did, when he was already ill.'

'I must go to him, Nancy. I need to attend to his comfort.'

'Ain't nothing you can do for him, Miss. He must stay abed for days, Mr Betts says. Safer for him, the way this ship is rolling about.'

'Oh, how I wish there was a proper medical man aboard. Please thank Mr Betts for me, and ask him to visit me when he can, so that I can learn more.'

'Yes, Miss, I'll do that.'

It's an anxious time, all over again, as I try to cope with my own illness while worrying about John. Finally, the seas calm down and everyone is able to leave their beds in time to see a group of whales breaching. Even John is there, moving about cautiously but shaking off my concern.

At last, the mountains of Van Diemen's Land can be seen. Captain Todd is beaming with delight, happy that his calculations have been so accurate as to land us in the right spot. For many on board, like the Waterhouse family, this is the end of their journey, and we are all enraptured.

'I feel as a prisoner must, upon release from gaol,' John says. 'The sight of land, after so long imprisoned on this ship, is overwhelming.'

I'm pleased to hear that I'm not the only one who considers the ship a penitentiary.

'It's seeing trees again that delights me,' Mr Eggleston replies. 'They were marvellous indeed at Cape Town, but here they are such a swathe of foliage. The colours make me think of rural Italy with their olive trees.'

'Have you been?' I ask. 'I confess I have not visited that country.'

'No, neither have I,' says Mr Eggleston, 'but I have seen paintings of it.'

We've spent nearly five months at sea. After a long time on the Derwent River, opposed by contrary winds, we reach Hobart Town. It's an extensive and straggling town with wide streets and thinly scattered buildings. There are stone buildings and a few brick ones, but I'm surprised to see that most of the houses are made of timber. It gives the place a temporary feel. And there are no cobbled roadways. The paths are dusty and will likely turn to mud at the first rain.

We are welcomed by the Reverend Orton, the resident missionary. The Mission here is extensive, with a large and interesting congregation. They have a small but neat chapel, and the building of a beautiful spacious new one is in progress. Most of the *James* passengers are distributed to stay amongst various acquaintances, although a few of the men are obliged to return to the ship to sleep. John is invited to preach at the chapel on our first Sabbath. I give heartfelt thanks to the Lord for my safe passage and to be among these new friends.

We are to remain in Hobart Town for at least a month while onwards passage to New Zealand is arranged. The other ladies and I are happy to find some good shops with bountiful supplies at reasonable prices. I organise for the sailors to retrieve my hives from the *James*, and then for a carter to take them into the countryside. When they are fully opened, the bees fly in great clouds around the skeps, happy to be released from their prison of the ship. I know how they feel.

I check how well the skeps themselves have fared, after the long journey. Some of the cloaming, the protective layer of dung which waterproofs the straw, has cracked, but most is holding up well. They are very light, though. There isn't much honey left inside.

'Are there any badgers around here?' I ask the carter, Mr Batt.

'Badgers, Mistress? None I seen. There are some strange animals, mind you. I seen some spiky things, more big than an hedgehog, on Bruny Island. And roos, a course.'

'I'm just concerned about what might attack the hives,' I say. 'What about those koala bears I've heard about?'

'No'm, I seen none of them neither.'

There are large numbers of unusual flowering plants in the vicinity. One low-growing shrub has long slender white petals which fold back, and the bees seem to particularly enjoy visiting them.

'It's a grevillea, Mistress,' I am told. Batt says he was a farmhand before he was sent to the colony as a felon and takes an interest in the countryside. 'And look at them bees on the banksia.' It's another strange flower, rising from a necklace of leaves, with spikes of florets smothering a cucumber-shaped dome.

I hope that my bees will get their fill of nectar and pollen. They need to build up their energy before their final journey. What different plants will the bees encounter in New Zealand?

John is particularly well received in Hobart Town. Acquaintances try to persuade him to remain, rather than go on to New Zealand, and for a while I hope he'll agree.

'It is very pleasant here, with such a British sensibility,' he tells me. 'It would be rather like being at home.'

'Then will we stay?' I ask. 'They tell me you could be most useful here.'

'No, my heart tells me I must go to the heathen,' he says. 'So, in the name of my Master, to the heathen I will go. After all, the Conference has given me a task and expects me to fulfil it. And to tell you the truth, I feel thankful for it. There's not as much of a challenge in staying as in going.'

I'm not looking forward to a further journey by sea. I'm billeted with Mr and Mrs Heddestone, who've given me a neat little room with grapes growing around the window. It is wonderful to be able to sleep in a bed without fear of being thrown out. I spend my days visiting other missionaries or taking trips into the countryside. I think the country very romantic, with mountains close by. Even though it is midsummer, I can see snow still on the summit of Mount Wellington, and the weather changes easily from very hot to very cold.

A letter for Mr Waterhouse arrives from Mangungu with news that the Mission House in New Zealand has burnt down. This was where we were going to live, and I worry that we won't have somewhere to stay.

'Maybe this is another sign that we shouldn't proceed,' I tell John.

'Nonsense,' says John. 'I agree that we might experience some inconvenience when we first arrive, but the Lord will ensure we are accommodated.'

'But what if we need to live under canvas during winter? Or a mud hut? Your health is delicate, John. It might not stand the conditions.'

'Haven't you noticed how much the sea-voyage has revived me, Mary Anna? I feel ready for anything.'

Nancy and I visit the shops of Hobart Town again to purchase more blankets, oilskins, and canvas cloth. While I greatly admire John's fortitude, I do not entirely share it.

I try to imagine what a primitive life might entail whenever I take sojourns away from Town. One day we have a picnic dinner, another time we cook fish and make tea outdoors. It is pleasant enough in the sunshine of summer, but I still manage to take a cold. A further time I sprain my ankle. I fear I will be as poor at living roughly as I was at being a sailor.

John has seen some New Zealanders in attendance at one of the outdoor services and seeks out their company. He is determined to learn their language at the earliest opportunity. While still being men of colour, he tells me they are lighter in complexion than the natives of Van Diemen's Land, and more solidly built. John is also keen for me to learn how to speak the New Zealand language and returns from their meetings overflowing with useful information.

'They laughed when I tried to explain where we were headed for,' he says. 'The 'ng' sound is said almost as if the g is absent. So Mangungu is pronounced Mah-Nu-Nu, and the Hokianga River is Ho-Key-Ah-Na.'

After so many months of saying Man-Gun-Gu, I find it hard to train my tongue and mind, but I practise it under my breath so that I don't offend the natives when I meet them. John encourages me to start a small notebook of words, and to concentrate on the greetings and words of blessing.

As a group, we meet with His Excellence the Governor, Sir John Franklin. He is a kindly gentleman and keen to have us to dinner with his lady wife. When the day arrives, I am most interested to meet Lady Franklin. I've heard many stories from the locals of how interested and involved she is in the administration and works of the land. Lady Franklin is much younger than her husband, a second marriage for him. They are wonderful hosts, looking after thirty guests with sumptuous entertainment. I am not disappointed in Lady Franklin, who seems very interested in our missionary zeal, and even suggests she might visit New Zealand herself.

'I hear there are no snakes in New Zealand,' Lady Franklin says. 'I'm on a mission myself to rid Van Diemen's Land of the horrible beasts. I pay a shilling for each snakehead.'

'I'm not sure,' I say. 'I've yet to go there, of course.' Snakes have not entered my mind of things to fear about New Zealand. 'I hear you are a great traveller.'

'From my early years, I have been accustomed to exploring,' Lady Franklin agrees. 'There are so many wonders in the world, don't you find?'

I nod, while thinking that I don't have her fortitude or inclination to explore. I've heard that Lady Franklin and her companion, Sir John's niece Sophia Cracroft, climbed nearby Mount Wellington. The story I was told was that Lady Franklin's shoes were so worn by the time she reached the summit, she'd had to send for a new pair before she could make the descent.

Buoyed by Lady Franklin's sense of adventure, I resolve to make the most of venturing into a new country.

'Mah-Nu-Nu,' I whisper under my breath.

'What was that?' Lady Franklin asks.

'Mangungu,' I reply. 'It will be our new home.'

'You know,' she says, 'I believe I will come with you. I've heard so much about New Zealand. You'll have space on board your ship for me, won't you? I'll only need five or six companions. I'll have my things taken on board so that I'm ready to leave when you are.'

It is not my decision to make, but I wonder at her impulsiveness and willingness to venture into the unknown. It took me many months to fully commit to coming with John, even though I had indicated to him that I would.

'And Sir John?' I asked. 'Will he be coming with you?'

'I shouldn't think so,' Lady Franklin says. 'He has the important task of running the colony. I don't expect that they would spare him. No, I will report back to him on all that I discover.'

CHAPTER 6

After an enjoyable month ashore, it is time for us to embark once more upon *James*, which has undertaken to transport us on this last leg to the Hokianga River. We expect it to take only one week. This trip is a lot less crowded with stores and belongings, although Lady Franklin has loaded a considerable amount of supplies aboard.

My hives buzz angrily at being sealed back into their hogshead casks, not happy to leave this bountiful land. I hope there'll be a plentiful supply of flowers for them in their new home.

Lady Franklin arrives at the ship at the last minute. 'I am afraid I cannot go with you,' she tells us, as she directs her people to unload her possessions. 'My husband requires my presence here.' I take from that comment that her husband has forbidden her to go, which is surprising. I didn't think anyone had the fortitude to stand up to her.

I shake her hand in farewell and resist the urge to curtsy. She is the closest I will likely ever come to royalty.

'I will find a way to visit, Miss Bumby,' she tells me, *sotto voce*.

'I look forward to it,' I reply. It would at least be someone I'd met before, even if rather daunting a personage.

It is both familiar and disconcerting to be back aboard the same ship where I spent those long five months, without the companionship of so many of our number. I'll particularly miss the company of Mrs Waterhouse.

'My husband says he will see you again quite soon,' Mrs Waterhouse says, gripping my gloved hands between her own. 'He expects to visit New Zealand often, as part of his role as Superintendent.'

'My brother expects to travel back here also,' I say. 'But I think that once I have my feet on solid ground, I'll find it difficult to willingly board another ship. Already on the deck here, I can feel my constitution rebelling.'

'Oh, my dear, I wish you a smooth and safe journey,' Mrs Waterhouse says, tears misting her eyes.

Even after the break on shore of five weeks, my *mal de mer* comes on again with the waves. My consolation is that this voyage should be relatively short. When I am sufficiently well enough to go on deck, I spend my time searching the horizon for the coast and mountains of New Zealand.

'There it is,' says John, on our ninth day at sea.

'Those are just clouds, surely?' I ask. The horizon seems blanketed with them.

'No, indeed, remember from our previous journey that cloud often cloaks the land. That is where we will do our useful work,' John says. 'Although I admit it does look rather dark and dreary.'

'Where do we land?' I ask, as Captain Todd passes by.

'I'm sorry to tell you, Miss Bumby, but the current is pushing us southward. We will have to tack out to sea and wait for the breeze to change,' he says.

I feel as if I am being tortured, being in sight of land and yet unable to disembark. I watch as we sail away from a river mouth. On its left is a tumble of glistening golden sand, while on the right the cliffs are covered in low foliage. I am loath to let the vision of our destination vanish, but eventually they are too distant for me to see.

The next morning the winds change and a signal flag from the tree-clad hill tells us that it is safe to enter. The ship gets over the much-dreaded bar of the Hokianga River, with its turmoil of white-capped waves, and settles at anchor just inside the mouth. We stand on deck, taking in our surroundings. Compared to the ominous scene of the previous day, the sight of what lies before us is marvellous. On all sides, steep verdant hills descend to the waterline. The water is flat and reflects the colours of the sky. Behind us, waves crash, but on the ship we can only hear birdsong from the shore, chattering and piping across the water. There is no sign of any humans.

Captain Todd orders another flag raised, and a small boat is launched from the right-hand shore. It is the pilot, Captain Martin, come to guide us up the river, about twenty-five miles to Mangungu.

From further along the shore another boat rows out, with a short stout fellow aboard.

'Is this the *James*?' he calls, coming on.

'It is indeed,' Mr Creed shouts back.

'Wonderful! We've been expecting you.' He climbs aboard *James* with the assistance of some of the sailors, then introduces himself as Mr William Woon, one of the missionaries stationed near the river entrance. 'Welcome to New Zealand.' I think I detect a Cornish accent. There is much hand-shaking all around.

'We're here,' I say to myself, scarcely able to believe it. 'New Zealand. The Hokianga River. Just miles from Mah-Nu-Nu, our new home. What a strange contrast this is to Thirsk. Whoever would have believed I would find myself here.' I think how amazed Jane and Mother would have been and wonder whether they have been looking down on my voyage.

On our trip up the river the next day, I am startled to see a crowd of canoes streaming towards our ship. They are filled with natives, nearly naked apart from old coats and blankets. They have pigs, potatoes, melons and Indian corn, which they appear to be offering for sale. I stare at them, my first glimpse of these fearsome warriors. Many have green lips and markings on their faces, and the clamour of their conversation is unintelligible. I look to Mr Martin and am reassured that he doesn't show any concern.

The hills are deeply forested, a blanket of green with tufts of white flags waving a greeting to us newcomers. The green is dark and slightly forbidding, rather than the grey tones of the Van Diemen trees or the paler yellow oaks and alders of England. I look at the steepness of the hills and wonder how anyone finds a path through the wilderness.

Then, like an oasis of civilisation, we come to a cluster of buildings and a small chapel, perched on a promontory on our right. From the ship I can see the Mission House, standing white amongst the green of the trees and grass. Surrounding it are various other buildings, mostly in what I assume is the native style, simple huts of brown vegetation. Towards the bottom of the hill there are weeping willows and oaks, looking familiar but out of place with their yellowish tinge of leaf. It is our destination, reached at last.

When we arrive at Mangungu, we receive a joyous welcome from our fellow missionaries. Mr Hobbs and Mr Turner come on board and introductions start all over again.

'Reverend Bumby, we welcome you to New Zealand, as a brother to us in our faith, and to reinforce the efforts we have made thus far,' Mr Hobbs says. He has been one of the senior members of this establishment, and I imagine that he is grateful John has come to shoulder some of the load.

When we come ashore, I take note of the date. It is Tuesday 19th March 1839, and I intend never to leave this place. No matter what hardships lie ahead, I wish to remain on firm ground forevermore.

I am introduced to Mrs Hobbs and Mrs Turner, who have come down to the landing to welcome us.

'I was told that the Mission House had burnt down!' I tell them.

'Oh, I wish you would not remind me of that terrible night!' Mrs Turner says, turning to look up the hill.

'Indeed, it was burned,' Mrs Hobbs tells me. 'My husband rescued Mrs Turner and all the children, and her little boy Josiah was particularly fortunate to have been saved from the fire. Our own house also became ablaze and only narrowly saved from suffering the same fate.'

'But what's this house then? How did you get another built so quickly?' I ask.

'My husband has skills in carpentry and joinery, learned from his father, and he supervised the building of this new one. Our natives are sawyers of their own timber, which meant it wasn't a great expense. The inside is still unfinished, so in the meantime we will have to accommodate you in a whare.'

I'm not sure what Mrs Hobbs means by a far-ray, but I am sure I'll soon find out.

Both Mrs Hobbs and Mrs Turner look to be busy women, with numerous small and larger children gathered around their skirts. Mrs Turner seems fragile, which isn't to be wondered at. I count eight children dancing around her, one a nearly grown girl, and a babe in her arms.

Mrs Hobbs proves capable and practical. She tells me that she'll bring over a loaf of bread and some milk, and she has some native girls organised to get the fire going. It brings tears to my eyes to have someone so thoughtful looking after my needs after such a trip.

I trudge up the hill behind the women and children, my legs aching after so long without much exercise. I hope that Nancy or one of the native servants will be responsible for fetching water. Beside me, one of the little girls skips.

'Hello, I'm Miss Bumby,' I say. 'What's your name?'

'Pleased to meet you, Miss Bumby. I'm Emma Hobbs, and I'm ten. Have you come to live with us?'

'I have indeed. I've come with my brother, Mr Bumby.'

'I have a brother, Richard, and three sisters. I had another brother, but he died. It was too cold for him.'

'I'm sorry to hear that. How long have you been here?'

'I was born in New Zealand, and we lived here when I was little,' Emma says, 'but then we went to Tonga. We've been back at Mangungu just over a year.'

I admire the way this young child pronounces Mangungu. It is, as John explained, with the 'g' mostly silent.

'Well, Emma, it will be lovely having you as a neighbour,' I say. We pass a clump of grass-like leaves, and there is the white 'flag' of a flower, looking more like a brush, cream-coloured with tinges of pink. 'What do you call this?' I ask.

'That? Oh, that's toetoe,' Emma says.

'Toy-toy,' I repeat. I hope Emma might be a useful person to introduce me to the wonders of my new world.

Captain Todd declares that he is sad to lose his passengers, but very thankful that he's been able to deliver us all safely to our destinations. He puts on a grand dinner, followed by the firing of cannons and a great display of fireworks.

These delight the children and bring a large contingent of natives out to watch.

As we are admiring the lights, a ripple goes through the crowd as information is passed along.

'What's happening?' I ask Mrs Hobbs, who is standing nearby.

'It's Tamati Waka,' Mrs Hobbs says. 'He's come to see what is happening.'

'Is he friendly?' I ask. I hope he isn't one of the war-mongering natives.

'Oh yes,' Mrs Hobbs says. 'Tamati Waka is his baptismal name. It's their way of saying Thomas Walker, who is one of the great CMS supporters back home. His native name is Nene, and he's still known by that name quite often. My husband baptised him.'

'He's a Wesleyan then, despite his baptismal name?' I ask.

'Yes, and he's one of the great Ngāpuhi chiefs. His brother, Patuone, is likely to become Anglican. We suspect that they decided to have one of each, to keep both missionary societies happy.'

'Can we not use this Thomas Walker to bring his brother over to our side?' I ask.

'Why would we want to do that?' Mrs Hobbs says. 'The CMS are not our enemy. They've been particularly good to us. I stayed with them when I was expecting Emma, and again with Margaretta. We may not always agree with the particulars, but they are doing the same work as we are, of bring the word of our Lord to these people.'

Nene is now moving among the men, being introduced to John, Mr Creed, and Mr Ironside. I can see Mr Hobbs standing beside the old chief, presumably translating for him. From the way the New Zealander holds himself, I think that he probably understands English better than he is letting on.

'He has authority,' I tell Mrs Hobbs. 'You can tell by the way people treat him, and by his stance.'

'Yes. Here they call it mana. He's a kind man and looks out for us. He's always trying to bring peace to the area, even though he was apparently a fierce warrior in his youth. He fought with Hongi Hika, you know.'

'I think I've read about him,' I say. 'He was the one they called Shunghie, wasn't he? He came to England. Was he the one who was given a suit of armour by the King?'

'Yes, that's the one. Some people say that he caught the spirit of Napoleon as they passed by the island of St Helena on their way back to New Zealand. Certainly, when he got home he instigated some terrible wars against other tribes, and Nene joined him. But our influence has meant that Nene wishes for an end to the tribal fighting. He was one of the signatories to the Declaration of the Independence of New Zealand.'

'I'm sorry, he signed what?' I ask. I'm still watching him. He has a high forehead and wispy hair on top, with swirling tattoos on his cheeks. He is smiling broadly, obviously comfortable being around so many Englishmen.

'The Declaration? With the British Resident, James Busby? It happened while me and my family were in Tonga, so we weren't involved, but Nene and thirty other rangatira signed to say that the country belonged to the natives, and they were the only ones who could make laws here.'

'Why would the English want that?' I ask. 'And what's a rangatira?'

'It means a chief. And I believe our people were worried that the French would stake a claim to the country, if we didn't get alongside the natives. And of course, with the French come the Papists.'

'Is that likely, that the French would take control?'

'It was at the time. We have a French nobleman living here in the Hokianga, Baron de Thierry. He claims he was appointed Sovereign Chief by Hongi Hika back in England, which is ridiculous. Nene has given him a small amount of land to keep him happy. I'll have to take you to visit him sometime. He's a lovely man, actually, and more English than you would think. And last year the Frenchie Catholic Bishop Pompallier arrived. He's settled on the north side of the river, and we've lost a few souls to him.'

Nene has now made his way to us. He holds out his hand to Mrs Hobbs. 'Mihi Ropiha, how do you do?' he says.

'Nene, my dear friend. This is Miss Bumby, newly arrived from England.'

The man holds out his hand to me and smiles. I shake his hand.

'Mihi Pumipi,' he says. 'How do you do?'

I turn to Mrs Hobbs. 'I'm sorry, but what is he saying?'

'Our names, that's all. I'm Mihi Ropiha, and you are Mihi Pumipi. They find it hard to pronounce all the letters, so often their words are an approximation of ours.'

Nene points across at John, then back at me.

'My brother,' I say. 'How do you say brother?' I ask Mrs Hobbs.

'It's not that simple,' she says. 'It depends on the relationship. In your case, I think it's tungane.' She says this last word slowly, so that I can hear the three syllables, Tu-Na-Ne.

Nene smiles even more and nods his head. 'Good bye,' he says, and moves away. A small crowd of natives follow him.

'You're so clever with the language,' I tell Mrs Hobbs. 'I'm sure I will never be able to talk it.'

'You'll be surprised,' Mrs Hobbs says. 'We were away for five years, but it still came back. I'm not as good as my husband, of course, but it is a lovely language to speak. And they do appreciate it when we try.'

'I was introduced to Chief Nene tonight,' I tell John, as I pull on the fingers of my glove. We have retired to our hut for the night. Blankets have been hung to provide some privacy and create small bedrooms, and I'm preparing to enter my room. But first, I want to share my thoughts with my brother.

'Who?' asks John. 'Oh, you mean Tamati Waka? Yes, Mr Turner introduced me to him while Mr Hobbs interpreted. Mr Turner said I was the father of the missionaries.'

'The father! Whatever did he mean?'

'Just that I was the Superintendent. But Tamati Waka responded that I was but a boy, but perhaps I had the heart of a father.'

'He seems like a wise man. I saw that he rubbed noses with Mr Hobbs and Mr Turner, but not with you or the other new arrivals from our ship. And he certainly didn't try it with we ladies. It seems a strange custom.'

'I've been asking about it,' John says, sitting on a chair to take off his boots. 'It's called a hongi.'

'As in their name for John?' I ask. I hadn't expected New Zealand words to be in such use by the English.

'No, that's Hone. Can you hear the difference? Hone and Hongi? This has that N G spelling that gives a more nasal intonation.'

I can't hear the difference, but I spell the word out in my head. 'So, it's hongi, as in Hongi Hika the old warrior?'

'I hadn't thought of that. But yes, I think so.'

'Why do they do it? It seems disgusting to touch noses,' I say.

'The way Mr Hobbs explains it, the idea is to share the air you breathe with the other person. So, they are saying that they are part of you, and you are part of them. That we all belong as one.'

I have my gloves off now and sit down to unbutton my boots. 'That doesn't sound so bad. Why don't they do it with us?'

'Apparently, they've learnt that the English aren't always comfortable about it. Sometimes we've been suspicious that the New Zealanders' will stab us in the side, while our eyes are locked. Or steal from our pockets. Something that happens a lot more often in England than here, I imagine.'

'It does feel safer to remain at a handshake's distance,' I say.

'I intend to hongi as often as I can,' John says. 'It's a show of faith and goodwill, to accept their customs and language. I'm certainly grateful to Mr Hobbs for his interpretations, but I want to be able to communicate to the natives directly, as soon as I'm able.'

'It's a strange place, isn't it?' I comment. 'Not at all like I imagined.'

'Isn't it wonderful?' John replies. 'So wild and untamed. And the people are fierce and intelligent. The challenge of bringing our Lord to them thrills me to the core, and yet I wonder whether I am ready for the task ahead.'

I rise, swaying slightly from the remnant effects of the sea voyage, and touch John on the arm. 'I'm sure you are, John. You have the heart of a father, after all. This is why you've done so much preparation. And I'm here. We'll do this together.'

'Thank you, Mary Anna.' John kisses me lightly on the forehead. 'Get you off to your bed. There is another busy day ahead of us tomorrow.'

I am outside the yet unfinished Mission House, directing where our luggage should go. Furniture and stores are piled into the dry corner of a room which is almost complete, while more immediate needs are placed in the little V-framed raupo hut with its dirt floor, the one Mrs Hobbs calls a whare. An upside-down V, I reflect. Even the poorest people back home have more substantial dwellings.

I spy Mr Hobbs by my hives. He looks as if he is about to pick one up. I hurry over to him and place a restraining hand on his arm.

'Please,' I say. 'Leave that. They're my skeps, and they need to stay out here, slightly away from the buildings.'

'I beg your pardon?' Mr Hobbs says, startled at my touch. 'What are they?'

'My bees,' I say. 'I'll let them settle down overnight, and then they can explore their environs tomorrow.'

'You brought bees?' Mr Hobbs says. 'All the way from Home? Whatever for?'

'My brother likes honey,' I say. 'And I enjoy beekeeping. I thought it would be a good thing to do.'

'Miss Bumby's bees,' Mr Hobbs says with a smirk. 'Are you sure your name isn't Miss Humblebee?'

It isn't the first time such a comment has been made, but it still irks me.

'Bumby is an old and honourable name in Yorkshire,' I say. 'And they are honeybees, not humblebees.'

'I beg your pardon, ma'am,' Mr Hobbs says, and bows slightly as he moves away.

I frown as I watch him. He has an officious attitude that doesn't fit well with me. My joy at being back on land is ebbing, as I wonder how I'll deal with the other people here. It looks as if we are going to be thrown into each other's company a great deal. And that hasn't been the best start, although I'm not sure why.

I seek out John.

'I understand that we're being given the Mission House to live in,' I say. 'Yet Mr Turner and his family lived in the previous one. And Mr Hobbs and his family are crowded into that small three-roomed whare and have been there for over a year.'

'It's due to my position,' John says. 'The Society appointed me as Superintendent of the New Zealand Missions. You could hardly expect me to live in inferior accommodation to my subordinates, could you?'

'But the Turners?'

'I believe they're considering returning to the colonies,' John says. 'The weather in Sydney might suit Mrs Turner's health better.'

I begin to worry about how severe the weather is here in winter. I thought it was supposed to be milder than home. Yet Mrs Turner is obviously unwell, and young Emma Hobbs mentioned that her brother died because of the cold. I hope I've packed sufficient blankets. How do the natives manage in winter, with their scanty clothing? Maybe they have bearskins which they don as the temperatures fall.

I look over at the forest, not very distant from our houses. Apart from the toy-toy, it is all green. Where are the hedgerows of dear old England? Where are the meadows with clover and cornflowers, daisies and poppies? What will my bees exist on here? No wonder there aren't any hives. They have nothing in this country for bees to eat. I can always continue feeding them sugar, but what will they use instead of pollen? Without the beautifully coloured balls of pollen, the bees won't be able to feed their young grubs.

I should have thought to bring some gorse seeds. Their flowers are packed with pollen, and they flower for many months of the year. They make wonderful hedges as well. Perhaps I can ask one of my friends to send seeds, although by the time they arrive, the bees will be dead.

I shake my head to clear it. Tomorrow I will release the bees and allow them to explore. If they can survive, they'll find a way. This might not be the promised land of milk and honey, as in *Exodus*, but while we are alive, we'll work to make their best of it.

Miss Bumby's Mission

CHAPTER 7

I don my veil and my thickest gloves. I don't expect the bees to attack me, but it's best to be safe.

I have an audience. The children have never seen a beehive before, nor have any of the natives. I've had the skeps moved to the back of the churchyard, near the fence, and have lifted them from their protective barrels onto a base of rocks.

Mrs Hobbs and Mrs Turner are encouraging their children to stand back.

'They are like flies,' Mrs Hobbs is saying, 'but they bite. If you don't move over here by me, they might hurt you.'

'I don't think they will,' I say, 'but it is best to do as your mother says. They won't intend to injure you, but they will want to protect their home and their children.'

I take the cork from the opening of the first hive and stand back, waiting to see what will happen. One bee pokes her head out, stands still for a moment, then walks further out, climbing up the outside wall of the skep. Another two follow, and then a stream.

The first bee takes to the air, flies a small circle in front of the hive, and lands back where she started. Soon there is a crowd of bees, flying in ever-increasing circuits around the hive.

'Miss Bumby, what are they doing?' asks Emma, who has crept closer.

'They are memorising what their house looks like,' I say. 'Soon they will fly away to look for food, and they want to remember where to come home.'

'Miss Bumby, there's a dead bird in the woods behind my house. They could eat that,' suggests Richard, Emma's younger brother.

'Thank you, Richard. That's very kind of you,' I say. 'But bees don't eat animals. They only eat flowers.'

'But I like flowers,' says another Hobbs child. 'They're pretty. I don't want them to eat flowers.'

'Don't worry. They only have a drink from the middle of a flower. They don't chew on the petals. In fact, a lot of flowers like having bees come to visit them. The bees help the flowers to make seeds, so you can have more flowers the next year.'

'Ah! Aue!' One of the native men is dancing around, holding his finger out. A bee has landed on it, and he's stroked its back in fascination. Of course, it's stung him, much to his surprise.

I mime sucking on my finger, hoping that he'll do the same and that it will take the stinger out. The onlookers draw back from the hives. One of the children starts waving his arms in front of him, screeching.

'Mrs Turner! If you could stop him making those movements, please!' I call over the commotion. Mrs Turner corners the child and leads him away. 'Sudden movements frighten the bees,' I explain when I can be heard again. 'It's best to stay calm and quiet. If a bee comes close to you, just walk away.'

'What if one lands on you?' asks Emma, who seems fascinated. She points at the man who's been stung, who is showing his swelling finger to his friends. 'Like with Ari?'

In response, I extend my gloved hand towards the hive and encourage one of the newly emerged bees to walk onto my finger. I hold her up to the watching crowd.

'If I just wait quietly, this bee will decide that I am uninteresting, and she will fly away.'

Obediently, the bee rises into the air and is lost among the others.

I move over to the other skep and start to loosen the cork. One of the New Zealanders calls out something. I look up, and then over at the Hobbs family.

'He says Miss Bumby is a brave wahine,' Emma says. 'He says he would like to learn how to work with bees like you.'

'Thank him for me, please Emma. What's a wahine? And I accept his offer of help. What's his name?'

'His name is Heni, although we sometimes call him Henry. And a wahine is a woman.'

I look over at the man. He is young and strong, standing proudly erect, knowing that we are talking about him. I nod at him. 'Thank you, Heni, I would be delighted to have you help me.'

I listen while Emma relays my message to Heni. What a remarkable child she is.

'Emma,' I say, when the girl has finished. 'Could you ask them to watch for the bees when they are in the forest? If they see a flower with a bee on it, perhaps they could wait until the bee has finished and then pick the flower for me?' It has occurred to me that this might be one way of finding out what flowers are around, if any.

'Where should they put the flowers, Miss Bumby?' Emma asks.

'I'll leave a vase here, near the hives,' I say. 'That way the bees can visit the flower again before it dies.'

'Can we pick flowers?' asks the child who'd been concerned about the bees eating them.

'You certainly can,' I say, hoping that the collection will be sufficient for the bees' needs.

The gathered group begin to disperse.

'Emma?' I call. 'Could you ask the man who got stung to come and see me at the Mission House, please?'

'He wants to know if he is going to be punished,' Emma translates.

'No! I want to give him something for the sting,' I say. 'I have some cider vinegar which should help reduce the pain and swelling. I'll put it on a piece of cloth, and he can wrap it around.'

I can see one of Emma's sisters whispering to her.

'Miss Bumby? My sister Marianne wonders whether she can help with the bees, also.'

I look at the child. She is sturdy, a little younger than Emma and less confident, but from the look on her face seems determined. I remember my own childhood, and the kindness of Mr Webster in teaching me how to work with the hives. The important thing is to be interested. Heni will be able to do the heavy work, and if Marianne is fluent in the language like Emma, then communication will become easier.

'I would be delighted for your assistance, Marianne. But you'd better get your parents approval first. Then you can visit me, and we'll find a veil that you can use next time we work the bees.'

I wish the circling bees a successful forage and walk back towards the Mission House to find a bandage for the man with the sting.

The first Sunday John and I spend in New Zealand is a revelation. A thousand or more natives arrive to see the new Superintendent and attend the church services. The chapel can't contain everyone, and it is obvious why the new building is underway. John is delighted to hear the service conducted in the native language, and even more determined to master it. In the afternoon, he addresses the English service, where a great number of our congregation are fellow countrymen. Some are from the nearby timber yards, some are loggers, and others have come from further down the river, where they trade in flax and pork.

Where possible, wives accompany their husbands to the service and are keen to meet me. There are not many Englishwomen in the country, and they value any opportunity to share stories and information. By the end of the day, my head is whirling with names and advice.

After a week in our funny little whare, Mr Hobbs comes to tell me that the Mission House is complete.

'I'm sorry that you weren't able to move in directly,' he tells me. 'Let me show you around.'

The Mission House is spacious, with rooms for both John and me and a spare for visitors. John also has a study, and there is a room for entertaining as well as the dining room. It looks as if it is going to be very comfortable, with a porch along the front, and dormer windows upstairs. It is made of wooden planks laid horizontally and painted white, while inside it is lined with more planks. Some of them look wet, and I reach out to touch the drips.

'Don't do that,' Mr Hobbs barks.

I wonder whether he is paying me back for my abruptness with the skeps. I look to him.

'That's sap from the raw kauri, because we didn't have time for the wood to dry properly before we put it in. It will harden in time, but for now it is particularly sticky.'

'Oh, cow-rie, I've heard of that,' I say. 'Those are the trees which are rafted down to Horeke, aren't they?'

'Yes, and our own boys have been felling them and working the sawpits here,' Mr Hobbs says. 'Our finances do not stretch to being able to purchase ready-sawn timber from Lieutenant McDonnell at Horeke.'

'This sap reminds me of the propolis my bees collect to seal up gaps in their hive,' I say. 'I believe propolis can be used for varnish. Is this cow-rie sap useful, in a similar way?'

'I've heard it is used in sealing wax,' Mr Hobbs says, 'but there's only a small amount available.' He pauses, then adds, 'So, are you ready to move in? I can send some of my boys to help move the chests from your whare.'

The house is delightfully spacious after the cramped quarters of *James* and the low-hanging eaves of the whare. I enjoy unpacking the chests and boxes and finding places to put my things. I've never lived in a new house before. It is so clean, although it smells of linseed oil and turpentine from the fresh paint, with resinous odours from the timber. I need to keep the windows open during the day to circulate the air. Otherwise, the smells give me a headache.

One of the native girls Mrs Hobbs has found for us, Hemaima, has worked for missionaries before and is a decent cook. Pork and potatoes are in plentiful supply, with Indian corn, broccoli and carrots. Gradually my appetite is returning, and my looking glass shows that the colour is coming back into my cheeks. My dresses are loose on me, but I decide against taking them in, hopeful I will return to my former strength.

When Mr and Mrs Turner lived in the previous Mission House, they'd started a vegetable garden. After the fire, it became overgrown and trampled by the workmen. However, I find the soil is still more friable in that area, and I'm keen to begin growing some of the seeds I've brought from home. I get one of the native workers to dig it over for me, ready to start.

'You'll need to get them to build a fence around the house,' Mrs Hobbs warns me. 'There are wild pigs around now, as well as goats, and they'll get in and make a mess of it if they can.'

'I want to get my summer crops in soon,' I say. 'It feels warm enough to get them started.'

Mrs Hobbs laughs. 'Oh, Miss Bumby, it's all different here. It may be March, but you must forget the months and concentrate on the seasons. We are coming into autumn now, but it won't be like any autumn you've ever known. The only trees which change colour and lose their leaves are those we've brought from home. The peach trees in particular are very fetching.'

'Don't the trees break their branches when it snows, if they retain their leaves?' I ask.

Mrs Hobbs pats me on the arm. 'Relax, dear Miss Bumby. It never snows here, nor even frost. It does rain though, and everything can become damp and uncomfortable. I suggest you encourage your men to bring in plenty of firewood, to keep your place cosy.'

There is such a lot to tell my friends when I write letters home. At least they won't find it strange if I mention getting ready for winter, because they'll be doing the same by the time the letters arrive.

Despite my fears, the bees find plenty to forage on. The onlookers take my plea for flowers to heart, and a collection of blooms is often waiting for me in the vase I've placed at a safe distance from the hives.

Not many of the flowers are familiar. There are a few roses plucked from one of the missionary's gardens. I hope the women aren't upset at having their flowers picked. I'm not sure whether it's the natives or the children who've taken them. Another flower looks like a broccoli which has gone to seed, with its simple yellow four-leafed cross of a blossom. There is the sprig of a small-leaved, purple-flowered plant, and when I raise it to my nose to sniff, I confirm it's rosemary.

But there are also unfamiliar flowers. Two have small white flowers, one with five elongated petals, raised stamens, and leaves which are serrated like a knife. The other has very small leaves, five rounded petals and a dark centre with inward-facing stamens.

There's a purple-flowered plant, darker in colour than the rosemary, with many inflorescences on each flower spike, and an unusual arrangement of leaves off the stem.

And most spectacular of all, there's a scarlet flower with thread-like petals, although I'm not sure whether they truly are petals. Perhaps they are stamens.

The bees keep returning from their trips, sometimes with pollen packed firmly on their hind legs, so I relax a little. Still, I'm interested to know what the plants are.

'I can ask my father if you like,' little Marianne Hobbs offers, when I ask her.

'No, don't worry him,' I say. I haven't forgiven his mocking of my name. 'Perhaps we could ask your mother.'

'Mother doesn't really know about plants, though. It's Father who is interested. You know those trees?'

'There are trees all around us, Marianne. Which ones do you mean?' I ask.

Marianne takes my hand and leads me to the doorway.

'Those ones,' she says. 'They're Father's special trees.'

I've already noticed the trees she's pointing out. They're about six feet tall, with whorls of leaves circling the trunk at intervals. The thin spiky leaves stand up from the stems. But most distinctive of all, the tops of the trees resemble a cross. They are most appropriate for a mission, and I expect that they've been planted for that purpose.

'What are they called?' I ask.

Marianne bites her lips and raises her eyebrows. 'Father's trees?' she says. 'He got them from his friend Mr Edgerley. He's the gardener for Lieutenant McDonnell.'

'I think you're right, then. I shall have to consult with your father.'

I keep a selection of the flowers on hand so that I can question Mr Hobbs when I next see him, which is sooner than I expect. Marianne has obviously told him his knowledge is needed.

'Miss Bumby? I believe you are interested in some of the New Zealand flora,' he says, standing at my open door.

'Mr Hobbs! It isn't urgent. I don't mean to interrupt your work,' I say, a bit flustered. Recovering slightly, I point behind him and ask about the trees we can see.

'Ah, those? They're my Norfolk Island Pines. They should grow to fifty feet, and they stay straight and true, just as we should do. They aren't affected by the wind like many other trees. Everyone who sees them thinks they are wonderful. I imagine one day we will be able to see these trees from the Heads, and sail towards them knowing that we are returning home.'

Then I bring him my samples. Some are already wilting, but he appears to recognise them all.

'This one is late in the season, but it is very common around here,' he says, indicating one of the white flowers. 'It's called mānuka in the native tongue. The tree doesn't get very tall, but it makes good firewood. In spring it is covered in these little flowers, sometimes so much that it appears to be snow. This other white flower is from the houhere, which is sometimes called lacebark.'

'And this pretty spiky red one?'

'That's a rata. It starts off as a vine, and then the trunk grows more sturdy and can support itself. The flower is very similar to that of another tree, the pōhutukawa, which puts on its display at Christmas.

'And this, ah, this,' he says, indicating the complicated purple inflorescence, 'this is *hebe speciosa*. First described here by one of our visitors, Mr Richard Cunningham, shortly before I left for the Friendly Isles. He was sadly lost in Australia a few years ago.'

'I'm sorry to hear it,' I say, impressed that Mr Hobbs knows so much. 'Is it a tree also?'

'No, just a shrub, but it can be a miracle of colour when fully in flower. Hebe was the Greek goddess of youth, but I've no idea why the plant was named after her.'

Mr Hobbs seems unlikely to have a classical education, so I'm astonished that he mentions this.

'I appreciate you helping me identify these,' I tell him awkwardly. 'It's bewildering to be surrounded by so many unfamiliar things. I expected the native people to be different from us, of course, but I didn't know there'd be so much forest, and none of the trees I know.'

'I'm surprised a lady like you is interested,' he replies. 'Usually, it is the flowers from home which are of more concern. Things like roses and hollyhocks.'

'Perhaps you don't know what I am like, then,' I say, stung by his assumptions.

'You're correct. You certainly are unusual, concerning yourself with botany like this.'

'Good day to you, sir.' And I shut the door on him. Just when I thought we were starting to connect, he puts his boot in his mouth. What a discourteous man.

The following Sabbath is Easter, the greatest celebration of the religious year. I've worked hard with my servants to prepare meals in advance, so that we can concentrate on the festivities and not do any physical labour during the day.

John gives a wonderful sermon to a congregation where a great number of Englishmen are again present. I listen avidly.

'I have learned, in whatsoever state I am, therewith to be content,' he says authoritatively.

I think back to our conversation after meeting Nene. Unless he learnt this in the last week, then he hasn't fully gained the ability to be content. But it is true that he seemed happier on the ship coming here, and in the nearly two weeks that we've been here, than he was in England. Generally, his health is better as well.

'Discontent is a grave where all God's mercies are buried,' he continues, saying the words slowly so that we can take in their import.

It is true, I think. God's mercies are always available, but when you are lost in misery, you also lose sight of what is good. I think back to our voyage. The times when I thought God had forsaken me. The continual dragging sickness that made throwing myself off the ship seem an alternative, despite the danger it would inflict on my soul. Yet here I am, having survived the torment, arrived in this strange and beautiful land. A land where I hope to contribute in a meaningful way.

I am roused from my reverie by everyone standing and starting to sing. It is the hymn *Christ the Lord is Risen Today*, one of my favourites. The soaring '*Alleluia*' at the end of each line makes my heart rejoice. The line '*Raise your joys and triumphs high*' reminds me of the force of John's preaching. Then the congregation moves on to

Love's redeeming work is done, with its lines '*Lives again our glorious King; where, O death, is now thy sting?*'

As I sit again, the idea of stings makes me think about Marianne and Heni's first visit to the hives yesterday.

Heni is a pipe-smoker, so I persuade him to blow smoke at the bee entrance. As the bees calm, Heni carefully inverts the skep and we look inside. The comb is dark, as I expect after having the bees walk over it for so long. There is nectar glistening in some of the hexagonal chambers, although a portion drips out when tipped. The bees haven't had enough time to ripen it into honey or to cap it with a layer of wax, sealing it in.

I briefly see the queen bee as she scurries away from the light, but I'm not quick enough to point out her larger, sleeker shape to the others. Considering the bees have only spent ten days on land, the future of the hives looks very promising.

'Look, Miss Bumby, the bees are holding hands!' Marianne exclaims.

I'm impressed with her observation. She's showing no fear, but taking it all in.

'That's right, Marianne. That's called festooning. They often do that when they're making new comb. It might be how they measure the distance between each layer of honeycomb.'

'How do they make the comb, Miss Bumby? And why do they make it in that funny shape?'

'They seem to be able to make the wax themselves,' I tell her. 'And that shape is called a hexagon. It means that there is no wasted space in the comb. It's nearly a circle, but circles would leave unused bits between them, while these hexagons are able to crowd into each other perfectly.'

'Clever,' she says. 'Clever little girls.'

Marianne receives a sting, not when she is at the hives but when she is taking her veil off. A bee becomes stuck in the fabric and attacks Marianne when she places her hand on it. She's stoic after the initial yelp of surprise and although she has tears in her eyes, she accepts the vinegar poultice with good grace. I'm surprised, especially when I consider how young she is. It doesn't seem to lessen her interest.

Nancy settles well into our new home. The servant girl seems to enjoy overseeing the natives and needs little instruction from me. In fact, she is learning more of the language than I am, due to her proximity to Hemaima and Mata. And those girls, in turn, are learning more English.

Mr Hobbs warns us not to teach the natives except in their own language.

'It's a matter of influence,' he tells John soon after we arrive. I'm listening in to their discussion, interested in learning more about how things are done here.

'I would have thought that the better the New Zealanders could communicate with us, the faster we could civilise them,' John replies.

'That would be true if we were the only ones coming to this country,' Mr Hobbs says. 'But there are all types of influence, especially in the Bay of Islands. Whalers, sealers, merchants. They spread their ungodly thoughts and habits among the natives and undo the good we're trying to achieve.'

'So that is why we are translating parts of the Gospel into their language?' John asks.

'It is indeed,' Mr Hobbs says.

I can't keep out of the conversation.

'Mr Hobbs, how many can read their own language?' I ask.

'Until now, very few,' he replies. 'That is why we have classes. They're eager to learn, and many of the chiefs send their sons to us to learn. Then they return to their tribes and pass that knowledge on.'

'And the girls?'

'We've had classes for girls too. Mrs White was useful for that when she was here. I'm not sure whether she's continued in their new residence.'

'I wonder whether I should do something similar,' I say.

'It would be difficult for you at present,' Mr Hobbs says, 'with your lack of understanding of their language.'

'And you have enough else to do, besides,' John agrees.

Then the men go back to their discussion on ways to bring the native people to the Lord.

Mrs Hobbs also approaches me to discuss teaching.

'My friend, Mrs White, set up a school for girls when she lived here,' Mrs Hobbs says. 'Girls like your Hemaima and Mata. They were taught to read and write, and also sewing and cooking skills. I wonder whether you'd like to restart it. I've been meaning to, but I've been so busy with the children.'

'Yes, that's something I believe I'd enjoy,' I reply, although I wonder when I'll find the time. I remember John's dismissal of the idea, and suddenly I'm determined that I'll find a way. How dare he think that all I'm good for is housekeeper work!

It's true that most of my hours are spent around the Mission House, instructing and supervising the work of the staff. They tend to waywardness, arriving at the House whenever it suits them, then disappearing for days on end. It aggravates me that I can't communicate with them easily, and I spend time each day with Hemaima and Mata, practising their language. The native women explain about the importance of family relationships and their funeral customs, and I begin to understand when they fail to arrive for work that it's because some distant relative has died.

Supplies are carefully watched, counted and guarded, especially because they are so difficult to replace. I'm happy to note that in general the natives are not light-fingered. Also, although they like to smoke and in fact are often paid in tobacco, they don't touch alcohol. There is a strong Temperance Society which was formed three years previously by Mrs White's husband William, and the owner of the Horeke shipyard, Lieutenant McDonnell, and its influence has spread throughout the Hokianga community.

I find that the New Zealanders are industrious, and I'm able to purchase meat, fish, onions, potatoes and fruit from their communities. They have a curious type of sweet potato they call kumara, which goes particularly well in pork stews.

I write a letter to my old cook, reassuring her that we have arrived and are eating well. Cook doesn't read, but she'll find someone to read her the letter's contents. She's moved back to Yorkshire to be closer to family. I miss her dour humour and feel grateful for all the thought she put into preparing me for this new life.

Cook's soap is carefully eked out for washing. The bright sunshine, clean water and lack of smog mean Monday's washing is blindingly white when it comes in from the line. I bury my nose in it, savouring the sun-warmth and the freshness. I'll have to start saving the wood ash to make lye, so that we can produce our own soap from all the lard we have. It's lucky the natives can supply us with a plenitude of pork, since pig fat is so much nicer for soap than tallow.

But Nancy isn't so happy. 'If only we'd a brought a mangle,' she complains each washing day.

'I know it's hard to wring clothes by hand,' I say. 'But look at how well our buttons are faring.'

''Tis hard to get all the water from the clothes,' she insists, as I help her twist a sheet.

When we left England, I had considered loading one on the cart. But they are so heavy and bulky. 'Maybe we can ask for one to be sent out,' I suggest.

Sugar is the most precious of commodities, especially at this time of year when the fruit is ripe and ready for preserving and jam-making. I'm trying to save enough sugar to supplement the bees until they are established, but I also want to make sure I have cakes and biscuits available to entertain visitors when they call. I wish I had honey to offer as well. Hopefully by this time next year, I'll have a harvest to share.

CHAPTER 8

I can tell that John is taking his responsibilities as Superintendent very seriously. Some of the other missionaries are older and vastly more experienced, yet John oversees their efforts. So, I'm not surprised when, only two weeks after we've arrived, he clears his throat at dinner and makes a pronouncement.

'It's been decided. We need to establish further stations to the south of this island,' he says, blinking hard.

'Oh, yes? How will you do that?' I ask, lowering my cutlery.

'Well, to begin with, Mr Hobbs and I are planning a trip to the Bay of Islands. And Captain Todd has decided to come with us.'

'Is he to take you on *James*?' I ask. The very idea of another sea voyage makes my heart beat faster. Not in a good way.

'No, there is a path of sorts,' John says. 'It will be quicker than taking a ship around the top of the island. We'll visit the Anglicans to discuss areas in which we might place new stations. And we'll enquire after a ship which might take us to those further regions.'

I can't imagine voluntarily boarding another ship so soon after making *terra firma*. I'm also not keen on John leaving me behind, but I have plenty to do in setting up the house. And maybe it would be an opportune time to think about a girls' school.

John, Mr Hobbs, and Captain Todd leave early on Saturday, in the company of three New Zealand boys, to act as their guides and porters. I stand on the porch to wave them goodbye. They look a strange trio: John with his long thin body, Mr Hobbs shorter and sturdier, and the captain with his loose limbs and rolling gait. I've pressed cold meats and baked potatoes on John, but he's refused to take much, saying that they'll be well-fed by their CMS hosts.

It's quiet in the house, even though John is often absent talking to the other men. But I'm startled when I hear knocking at the door.

'Miss Bumby, I wonder whether you would like to join me and the children for a meal after Mr Turner's service tomorrow?' Mrs Hobbs asks, when I greet her. 'It will save your girls from preparing dinner just for yourself.'

'That would be wonderful,' I say. I think of something else. 'Can I ask a favour, while you're here? Could you see whether Hemaima and Mata would sleep here overnight, while John is away? I know you are close by, but I think I would rest more comfortably if someone else was in the house, apart from just Nancy and me.'

'Certainly,' Mrs Hobbs says. She walks towards the kitchen with me. Our footsteps echo loudly on the bare floorboards. At the doorway she hesitates. 'Miss Bumby, there are very few women here at the Mission.'

'Yes,' I agree, wondering where this is leading. Is it to do with our proximity to so many natives?

'And we assist one another when we can,' Mrs Hobbs adds, watching my face. 'We become great friends.'

'I can see how that would happen,' I say. Mrs Hobbs has already proved herself thoughtful and considerate.

'I hope you don't think me presumptuous if I ask you to call me Jane?'

'Jane?' I ask. I clasp my hands to my chest. 'My dear late sister was Jane. A lovely name. I should be honoured. And you must call me Mary.'

Mrs Hobbs – Jane – laughs. 'Mary? One of my own sisters is Mary.'

'Then it was meant to be for us to be friends,' I say, putting a hand to Mrs Hobbs' arm. 'Come, Jane, and help me ask the girls for their company.'

When I go to dine with Jane and the children, I have questions for them.

'I've heard that the natives keep slaves,' I say. 'Are Hemaima and Mata slaves?'

'Gracious, no!' says Jane, serving me a piece of pie. 'All the servants we have here are quite highborn. They are being given the chance to learn our ways and will take those understandings back to their people.'

'But the New Zealanders do have slaves?' I ask.

'Yes, I believe so. Their slaves are from other tribes and have been captured in war. But they would have been lowly anyway, otherwise they would have been killed or ransomed. Is that right, Emma?'

'Yes, Mama.'

'But can we not make them give it up?' I ask. 'After all, slavery has been abhorrent to the English for many years now. How can we stand by and accept it here?'

'Mary, you misunderstand. We have no authority here, except for the moral authority given us by God. We cannot make the New Zealanders do anything at all.'

'We are making a difference though, Mama,' Emma puts in.

'We are, my dearest,' Jane says. 'The number of wars of retribution have decreased. And the instances of them eating their foes or their slaves is a lot less.'

Suddenly, my appetite for the pie, potato and beans has vanished.

'Anyway, to your original question on Hemaima and Mata. Are they suitable to your needs?'

'They seem nice clean girls, and they get on well with our Nancy. I'm not sure how much I should be paying them, though.' I feel that I've missed part of the process, not having selected them myself.

'Many of the natives are happy to work for clothing,' Jane says. 'And for tobacco.'

'I think I should be paying them as much as I pay Nancy,' I say. 'They get shelter and food, of course. Does five shillings a week sound suitable?'

'Mama, I want to go work for Miss Bumby,' little Margaretta says.

Everyone laughs.

'Shush, Retta,' Jane says. 'Miss Bumby already has enough servants. Now, finish up your meal, children. I have pudding ready for you.'

The four days that John is away go relatively quickly, with the company of Jane and Mrs Turner, who turns out to be Ann. While I have friends who I've spent time with back home, none ever lived in such close proximity, nor did they have so much information to share. The conversations I have with Jane and Ann are very different to Emma Sell's gossip about royalty and clothes. Even dear Sophia Hyde didn't swap recipes for pickles or give me gardening hints. Still, I miss my old friends and often wonder how their lives are progressing, back in smoky Birmingham. It will be many months before I can expect to get letters from them.

When John returns, he is full of talk of his adventures.

'The forest here, Mary Anna, is magnificent. Such a girth to the trees. I know we've seen them come down the river to the mill, but to observe them standing with their heads in the clouds is enough to inspire awe. Tree after tree. And then the vast tracts of fern, so lush. I shall have to get hardened to the walking, though. They might call them tracks, but you couldn't take a cart or carriage along them. No bridges and some particularly steep areas to climb.' He is pacing the room, too excited to sit despite his tiredness.

'I find I'm getting more used to walking up the hill,' I say, watching him from my seat. 'I don't have to stop and catch my breath, as I did when we first arrived.'

'It's a healthy place to live, I must say. The first place we arrived at was the church mission station of Waimate, where we got a wonderful welcome from the Reverend Richard Taylor. Mr Hobbs is a great friend of theirs, and he was invited to officiate at the service there, with one of their own ministers.'

'Yes, Mrs Hobbs was telling me she spent some time with the CMS when she was with child,' I say, nodding. I want to tell John how friendly I've become with Mrs Hobbs in his absence, but his focus is on his own experiences.

'And then on Monday, we dined with Mr Busby, who is the British Resident. A most accomplished and affable person. He's doing all he can to overcome the wickedness of the population, especially the Europeans.' He wags a finger in my direction, as if he expects me to disagree.

'Was he able to find you a ship for your explorations?' I ask, returning to my sewing.

'Sadly, no. And it seems that many of the natives are reluctant to travel with us. Their presence further south is not welcomed by other tribes, who still hold a grudge against the Ngāpuhi for the earlier years of war.'

I look up at him again. 'I'm not displeased about that. It will give you an opportunity to build up your strength and to learn more about the customs and language.

Oh, by the way, look what I found on my doorstep this morning.'

I put down my work and go to the meat safe, returning with two dead birds. I lay them out on the table and stroke their feathers. 'Aren't they beautiful? I'm not sure who left them for me. Hemaima says she will bake them. Look at how their feathers change colour when they're moved, from green to blue to purple.'

John prods one. 'They remind me of pigeons,' he says. 'I hope they taste like that.'

'Hemaima says they are kererū. Look at their tiny heads, compared to their fat bodies. And I wish I could get my washing as white as the feathers on their breast. And then the contrast with the red beaks and feet.' I lift one of the claws gently with a finger.

'It will make a nice change from pork and fish,' John says, nodding his approval. 'I think these might be the birds we saw when we were walking. They swooped over the forest with a great whooshing of wings. The sound was eerie.'

'Maybe I'll ask Hemaima to keep the feathers,' I say. 'I can include them with letters I send home to our friends. They're like peacock feathers, but without the evil eye.'

'Yes, which reminds me, I must write a letter to the Society, telling of our trip,' John says. 'You'll excuse me, won't you?'

After waiting so eagerly for his return, it is disappointing to have John withdraw so quickly. I still have plenty to tell him about what I've learned from the other women. It is aggravating that he's been the one away meeting new people, while I've always been more interested in having company. I hope that if he is forced to remain here at the mission station, we'll have visitors I can entertain.

I pick up the birds to return them to the meat safe, and then go to find Hemaima and Nancy. There are still a few peaches left on the tree and they might make a nice stuffing for the kererū.

The weather, so pleasant until now, has turned blustery with frequent showers. The servants are finding it difficult to get the laundry dry, rushing out to take in sheets as a squall blows through.

The trees around the Mission House change colour, not in the usual autumn fashion, but in producing lighter-coloured new growth at their tips. Around the Station, the grass grows long, while Mr Hobbs' apple and peach trees gradually lose their leaves.

I'm enjoying the cooler weather. I have more energy now and I'm sleeping better. I feel that I've almost fully recovered from the travails of the sea voyage. I start exploring the surrounding woods and wish that I'd been braver in the previous month. The shade of the trees is wonderfully cooling. I must remember that next summer when the sun is fierce.

The Mission House becomes busy with visitors. John gathers the missionaries there to discuss their plans and schemes, and I am kept occupied providing tea and cake. On other days, he accompanies his fellows on their circuits, occasionally staying overnight. During those times, when I am not needed, I visit Jane and Ann, helping as they teach their children, or devising new games and activities.

'Miss Bumby, tell us a story, please?' asks Emma one day as we are sitting around the fireplace. The rain is keeping us inside and the children are bored.

'What should the story be about?' I ask.

'A boy!' says Richard. With four sisters, he must feel outnumbered.

'Alright,' I say. 'This is a long story, so I can't tell you all of it today. It's about an orphan boy back in England.'

'What's an orphan?' Margaretta asks.

I sink back in my chair. It is going to be a challenge to tell this story to the children. They've never been to London and seen the dirty streets and poor living conditions. How can I make them appreciate what it is like living as a lost child there, when they live in such an isolated paradise?

Jane gives me an appreciative glance and leaves the room.

I start on the tale of the little orphan boy Oliver, hoping that I can remember enough of Mr Boz's story to satisfy my listeners.

A couple of hours later, little Phoebe is asleep, and Richard is starting to lose concentration. The older girls are enthralled, however.

'That's where we leave it for today,' I announce.

'No, Miss Bumby, tell us more!' Emma says.

'Yes, yes, tell us more. Please,' Margaretta begs.

'Not today,' I say. 'My throat is getting dry from so much talking. Why don't we try some string games instead?' I remember playing them with my sister, slipping our fingers through the threads and making fanciful shapes and patterns.

'How do you do that?' Emma asks.

'First, we need some pieces of string,' I say. 'Or wool would do as well. No, not from your mother's knitting! Yes, those ends of string will do nicely. Then we join them, like so, to make a loop. See?'

I get a loop ready for each child, then one for myself, hoping I can remember the moves.

'Are you ready? First, we are going to make a star.'

I show them how to hold loops with their fingers, demonstrating slowly. Phoebe has woken again and is desperate to be involved, but she lacks the coordination to make the moves. I sit her on my lap and use Phoebe's string to show the others, keeping the little girl's hands and fingers still while I pick up each loop.

'I've got it!' Emma exclaimed. 'Oh, so have you, Marianne.'

'Me too!' says Margaretta. 'See Emma, I've got a star here.'

'Come here, Richard,' I say. 'You've nearly finished. There, now you have it. Alright, take the string from your fingers and make it back into a circle. Now try again, from memory.'

When everyone is consistently making stars, I move them on to learning about a spiderweb. We are in the middle of excited chatter when Jane reappears. She beams.

'Mary, you have earned my forever gratitude!' she says. 'You must stay for dinner. Emma, would you go over to Miss Bumby's house and let Nancy know, please?'

'Mama, look what I can do,' Phoebe says, showing her mother a tangle of string.

'Lovely, Phoebe. Oh Mary, I'd forgotten about string games. I used to play them all the time. After dinner, we must show you one of the games the native children play, with sticks. It's noisy, but it's a lot of fun.'

CHAPTER 9

I am disappointed when John finally finds a ship to take him on his journey of exploration. I know that it isn't possible to keep him at the mission forever, but the idea of him travelling so far away worries me.

The ship is the *Hokianga*, appropriately, and she is a small vessel which moves supplies around the coastline. She is due to visit Kawhia in a few weeks, and her captain is agreeable to going the long way around, by the East Coast and Cook's Strait then up the west coast past Taranaki, for a reasonable fee.

'I'm hesitant to employ a ship,' John tells me that night. 'It's quite an expense, and we'll be absent from here for a long time. But it will open so much more territory to us for our endeavours.'

'Who'll go with you?' I ask. I would like to see more of the country, but the thought of another ship journey isn't at all enticing.

'Just Mr Hobbs and some of the native men. I've managed to get Mr Turner to agree to stay here until we return. I know he's anxious to return to the colonies, but I still need Mr Hobbs to be my interpreter.'

The month he's spent in Mangungu has seen John's grasp of the language much improved, but he isn't yet able to communicate all the nuances of his beliefs.

'I thought you said that the natives weren't interested in venturing south?' I don't want John inadvertently causing another war by taking Ngāpuhi men into enemy territory.

'Our excellent Tamati Waka has found us a solution,' John says. 'He's given us twenty of his slaves. They're originally from further south, and they're happy to return to their brethren and friends.'

'Nene?' I ask. 'How kind of him.'

'Even better news; many of them have experienced the grace of God because of their work among our missionaries, so they're eager to tell their kin about the greatness of the Lord.' John is grinning now, satisfied that he has found a solution that will benefit all.

'When do you leave?' I ask, thinking of what I need to do to prepare him for the trip. 'I thought the *Hokianga* was about to depart?' At dinner yesterday, Captain Barker mentioned that he'd nearly finished loading his supplies of flax. The chief engineer on the ship is Mr Thomas Marsden. I wonder if his famous Uncle Samuel has influenced his own beliefs, and caused him to pressure the captain into supporting our cause.

'Next Monday,' John says. 'We'll walk over the hills to the Bay of Islands and meet the ship there. That will give them time to unload.'

It's interesting to hear John talk about the journey to the busy port town as if it is of no consequence. I think back to the first time he walked it, only a month previous, and how he returned so exhausted. It shows how vigorous he's become. All this exercise means he is sleeping better, and he doesn't show his earlier propensity to abstain from meals.

I feel that I've settled into my new life as well. I'm not scared that I'm going to be murdered in my bed and eaten. All the natives I've met so far have been friendly and welcoming, and it's hard to credit the stories I hear from Mrs Turner of her early days in New Zealand. I really can't picture Ann fleeing in the middle of the night with her small children, in fear of her life with only the clothes on their backs. If she was so miserable, why then did she return to this life? Although that is the problem with marriage – you are required to follow your husband, no matter your own thoughts.

The day arrives when John and Mr Hobbs are to leave. John bids me his usual formal farewell, no doubt hoping that I won't make a fuss. So often I wish he was less constrained, so that I can express my feelings more honestly. If only I was young and could cling around his neck, like little Emma does to her father.

'Don't go, Papa, we need you here,' Emma pleads. She looks wildly around. 'Don't we, Marianne, Margaretta?'

'Come now, Emma,' Mrs Hobbs says, unwinding the girl's arms. 'We must allow Father to do his work, and trust in the Lord to return him to us.'

I am surprised to see Mr Hobbs crouch back down in front to Emma. 'I promise, if God is willing, I will return to you as soon as I may. Help your mother, do your schoolwork, and write me a letter. Can I rely on you?'

Emma looks tearfully at her father. 'You may,' she says. 'I will do my duty.'

The days are becoming cooler and shorter as May wears on. Some mornings when I wake and look out the window, I can see a mist covering the river, with the hills rising footless from the smoky haze.

Worried that my bees won't have enough honey to last through the winter, I set out dishes of sugar water for them. I hope that it won't cause robbing, where one hive takes to raiding the other for supplies. Sometimes at home I found that a weaker hive was completely destroyed by the actions of another.

'Miss Bumby, Miss Bumby,' I hear Emma Hobbs calling one afternoon. 'Come and look.'

Marianne makes it a habit to visit the churchyard each day to check what is happening with the bees. This time she's seen something quite different. She's told her sister, who then fetches me.

We stop at a distance and watch as five black-feathered birds squabble over the dishes of sugar water. They are aggressive, chasing one another off before dipping their beaks in the water and throwing back their heads. At their necks are pompoms of white.

'Parson birds,' I mutter. 'How strange that they like sugar.' I've noticed them before but expected that such large birds would eat insects or skinks.

'We call them tūī,' Emma says. 'Why do you call them parsons?' She shifts on the rock she's found to sit on.

'The Anglican ministers back home have started wearing white collars, which only show at their neck,' I say. 'These birds with their white tufts look like ministers, or parsons.'

'Interesting,' says Emma. 'You know, I'm going to marry a minister. He'll have to be a Wesleyan of course.'

'Of course,' I say, startled. I stare at the girl. 'But Emma, you're only ten, aren't you? Isn't it a bit early to be thinking of marriage?'

'You have to know what you want, don't you?' Emma says. 'And the best people are Wesleyan missionaries. Mr Turner, and Mr Woon, and Mr Wallis. And Mr Bumby, of course. What about you, Marianne?'

'Yes,' Marianne says, nodding. 'I'll marry a minister.'

'What about you, Miss Bumby? Who are you going to marry?'

The honesty of children, I think. No-one ever asks me that sort of question these days.

'I don't think I will marry,' I say, folding my hands into my lap. 'I'm too old for that now.'

'You're not as old as Mama,' says Emma. 'And she's still having babies. She's going to have another one. I heard her telling Papa.'

A few too many truths are coming out of this conversation. 'Oh, look at that little bird,' I say, pointing. 'He's not a parson bird, but he's interested in the sugar as well.'

'No, that greeny bird is a korimako. The tūī will try to scare him away because they're bullies. Look, he's waiting there for when they get full, or fighting so much they don't notice him.' Emma is properly distracted now, much to my relief.

Marianne whispers something to Emma. She is less shy when Emma isn't around, I notice.

'That's right,' Emma says. 'Both these birds are beautiful singers as well. Especially the korimako. People say he sounds like small bells ringing in the forest. The tūī are good at copying, so sometimes you think they're another bird.'

One of the tūī has decided he's had enough. He takes off, flying quite close to our heads. His wings beat a whish-whish sound in the air.

'Well, Marianne, what are we going to do about feeding the bees?' I ask, rising slowly to my feet. 'These greedy birds are going to use up all my sugar.'

'Hide it?' suggests Marianne, standing as well.

I consider the idea. 'Yes,' I say. 'We should be able to do something like that. After all, the bees can get into smaller spaces than these birds can. Let's go back to my house and find something that will work.'

There is a spate of fine weather, wonderful for outdoor activities. Jane organises an expedition, inviting Ann and me to join her and the children. We set off with a picnic meal carried by several of the native men.

'I've heard that there is a cave nearby,' Jane tells us. 'It should be interesting to explore.'

The older children rush ahead, causing some concern as their figures get swallowed up in the dense forest.

'Come back,' Jane calls. 'Emma, Marianne! Ann and Martha! Come back.'

When they return to the party, she scolds them.

'You can't go out of sight,' she says. 'It would be easy to get lost.'

'But Mama, we want to be the first to see it!' Emma says.

'And you shall,' Jane says. 'But only go as far as we can see you.'

Close to where she thinks the cave is, the native men start muttering among themselves and falling behind.

'What's the matter?' I ask, looking back at them. 'They don't look as if they want to come with us.'

'Here, could you hold Phoebe for me?' Jane asks, and she goes back to talk to the men.

When she comes back, she is looking worried.

'What is it?' Ann asks.

'I think they are saying that this place is tapu,' Jane explains.

'Tapu! But I'm sure Nathaniel has been here. And Mr White.'

'What's taboo mean?' I ask, looking from one to the other. They both seem uncertain about whether to continue.

'Tapu means forbidden or sacred, in their old ways,' Jane explains. 'Children!' she calls. 'Come back again. We're having our picnic here.'

'But Mama, what about the cave?' Emma protests when they are back.

'It's something we have to discuss,' Jane says. 'Let me think on it.'

It is a rather chilly place to sit and eat our meal of cold meat and pastries, out of the sun. I can feel the cool of the stone I am sitting on seep into my nether regions. We finish up, and then Jane sends the men home with the nearly empty baskets.

'What do you think, Ann?' she asks. 'Do you consider we should continue to the caves?'

Ann pauses, biting at her lip. 'Well, the children are very keen,' she says.

'Yes, but it won't hurt them to be disappointed.'

'And it is but a pagan understanding that makes a place tapu.'

Jane nods. 'I agree,' she says. 'We will trust that God will look over our adventures, rather than sinking to their level of superstition. Children, we go onward. But not too fast, please.'

I'm not sure whether it is due to the break from walking, the food, or my need to warm up again, but it doesn't seem much longer before we arrive at a large rock with a cavity in it.

'A cave, Mama!' Richard calls. 'Can I go in?'

'Let one of the ladies go first, please Richard,' his mother says. 'Mary, would you like to take us in?'

I'm not sure that I want to lead, but it feels as if I'm being given an honoured position. I duck my head and stare into the darkness, seeing only dimly-lit rocks. Carefully, I tread from boulder to boulder, my hands outstretched for balance. Close behind me come the children.

The place smells damp and green. I think I can hear the slight trickle of a stream, although it is dry underfoot. Gradually my eyes adjust to the faint light, and I can see further into the dingy cavern. The sounds from outside become muted, and I can hear the clatter of footsteps on rock and my own more rapid breathing.

Suddenly I see what is at the edges of the walls. I give an unanticipated shriek.

'Mary! Are you safe?' comes Jane's call from the entrance.

'Yes,' I reply. 'Only, now I understand why this place might have been tabooed.'

Set on natural platforms and rocks on each side, skulls and bones are resting. Not as full skeletons in a position of repose but gathered in groups as if placed there deliberately, long after death. An ossuary. I have heard of such, usually in association with papist practices.

The children have caught up with me now. I wonder whether I should be allowing them to see this, but it is too late.

'A crypt!' Emma says. 'How wonderful.'

'Are they people?' young Nathaniel asks. 'How did their bones get here? Did they get dug up?'

'Look at their eye holes,' Margaretta says. 'Can I touch one?'

Jane has joined us, leaving Phoebe at the entrance with Ann and her younger ones.

'No, we can't touch them,' she says. 'They were people, and we must show them as much respect in death as we would in life.'

'Why don't our native men want to come and see them?' Emma asks. 'We go and visit George in the graveyard all the time.'

'We have different beliefs of what happens after death,' Jane explains. 'They may think it's better to leave their loved ones to lie here in peace.'

I think of my own precious Jane, lying in her grave so many thousands of miles away. Is she in peace? Does she know that I have ceased visiting? Or is she aware that I continue to hold a special place for her in my heart?

'This shows,' I say slowly, looking around, 'that they may be pagans, but they still have respect for the dead. Even if they do eat their enemies, the bones of their loved ones are cared for.'

'Yes,' says Jane. 'It does give us hope that we can bring them to the Light. Their beliefs aren't so far removed from ours. Now, children, let us leave these people to their quiet.'

The following week Jane takes me to visit the Baron de Thierry, who lives nearby on some land Tamati Waka, Nene, has granted him.

'The Baron arrived here two years ago,' Jane tells me. 'He claims he's the Sovereign Chief of New Zealand, and that he paid Hongi Hika and Waikato for land when they were in England with Mr Kendall. Nene and his brother Patuone dispute that, but Nene let him have this small portion instead, which the Baron has named Mount Isabel.'

'Nene really is a peacemaker, isn't he? How wonderful that he's converted to Christianity,' I say. 'I can hardly believe that he used to be a warrior.'

'He's been a good friend to us, right from when we first arrived,' Jane says. 'I'm surprised that he agreed to be baptised as a Wesleyan, because he's also good friends with Mr Henry Williams of the CMS. Still, it's a wonderful thing to have such a powerful leader showing the way.'

We are treated with great hospitality by the Baron de Thierry, although his house is little more than a shack. They don't even have proper furniture, just cases to sit upon, while he occupies the top of a brandy barrel.

Hammocks are slung in the corners. His wife Emily, a petite woman, was brought up in Oxford. She is the well-educated daughter of an Archdeacon and has travelled with her husband and their five children through America and Panama, the Pacific Islands and Australia.

The Baron himself is very cultured and grand, with an engaging French accent. He is musical and plays us some tunes on his violin. He apologises for not being able to entertain us in a better manner.

'I had a large contingent of supporters when I arrived,' he tells us, 'But they have abandoned me. That McDonnell, he betrayed me.' He gives a Gallic shrug. 'I am in correspondence with the government of France, and when they lay claim to this country, I expect they will appoint me as Governor.'

His servant comes in with the tea, and he introduces her as Mrs Margaret Neilson. She is a tiny Scotswoman with a bustling manner.

'We couldn't do without our Margaret, could we Emily?' the Baron asks his wife. 'Myself, I am like a King Bee, with my subjects swarming around me.'

'I beg to differ,' I say. 'I know something about bees, as I keep two hives myself. They are like our own current society, with a Queen and no King.'

'Is that true? Very sensible indeed, then. I always think that women are indispensable in keeping society civilised. But what then of the male bees?'

'They are known as drones. They serve no useful purpose in the hive except to keep the Queen in child,' I say.

'Is that so? I never knew,' the Baron says. 'I should like to have bees myself, to learn more of this Utopia.'

Their daughter bounds in. A girl of about ten, she has curly dark hair and a sharp face.

'My Princess,' the Baron says. 'Isabel, let me present to you Mrs Hobbs and Miss Bumby.'

Jane talks to Isabel about her horse, which appears to be her boon companion. However, I find the child's manner to be haughty and selfish. Trying to be charitable, I wonder whether it is because Isabel is so isolated here, with no other girl children apart from the natives.

I feel great admiration for the Baron's attitude. He speaks as if he is destined for great things, although they are now living in such reduced circumstances. Their house must be draughty when the wind is up. At the front, a tree has been converted into a flagpole and a colourful crimson and azure flag flutters valiantly from it.

'What a remarkable person the Baron is,' I observe to Jane on our way home. 'Although, do you think the French will succeed in claiming New Zealand? It might badly affect our work if they do, and the Papists then flood in.'

'There certainly seem to be a greater number of French ships around the coasts,' Jane replies. 'And my husband tells me that Bishop Pompallier has unfortunately been very active and successful since he arrived last year. There was some upset and danger when he first arrived, and the Baron came to his defence.'

'Why did the Baron offer his territory to France? After all, I understand that he was raised in England and educated there. Didn't he say that he'd been to Cambridge?'

'I've heard the English don't believe he has the rights he claims. And although his parents were French and at Versailles before the Revolution, he was born in the Netherlands. I hear he's also approached the Hollanders with his offer. It seems the French are his third choice.'

'I wonder how long he'll stay in the Hokianga,' I muse. 'It doesn't seem grand enough for such a man. I would have thought he'd be better off in the Bay of Islands, closer to the capital, Kororareka.'

'I feel sorry for his wife. It sounds as if he doesn't stay long in any one place,' Jane says. 'It must be hard to raise your children when you are always moving.'

I hope that I can settle in this strange country. Each day, I find myself more accustomed to its ways. I wouldn't like to uproot myself as the Baroness appears to have done so frequently. Again, I think about how the lot of a married woman must be subservient to the wishes of her husband, and I feel grateful to be unwed.

CHAPTER 10

It is the middle of winter, with the bees only emerging from their cosy skeps on calm sunny days. It doesn't snow, but the frequent rain keeps everything cold. I need the fire going constantly to heat the house.

It is now apparent that Jane is again with child. That will make it their sixth living child. Not as great a number as the Turner's ten children, the complement of them full whenever the older boys return from school in Waimate. I can't understand how they cope. I've also learnt that Ann lost two children in infancy. No wonder the woman is constantly unwell; what a toll childbearing must take on her health.

Mr Turner is determined to leave as soon as John and Mr Hobbs return. His lungs are not enjoying the cool moist air of Mangungu, and he's been promised a position in Sydney.

After John has been away for eleven long weeks, finally a letter arrives from him. I open it with trembling hands, but it is proof that he is still alive – or at least was alive, at the time of its writing. It is in John's careful hand.

Kawhia, 18 July 1839

My dearest Mary Anna,
We have finally bid farewell to our vessel with no regrets, having been warmly welcomed here at the Mission House by Mr and Mrs Whiteley. We have had so many adventures that I can write but a portion of them.

Our journey thus far has not been without perils and privations, but also many mercies, thanks to the Lord's blessings on our work.

We were delayed several days in leaving the Bay of Islands, due to inclement weather. The seas continued heavy on our journey down the east side of the island,

and I was thankful that you were not with us to suffer as so many of us did. We passed a volcanic island which abounded with fine sulphur, that the natives name as Whakaari. They told us that Maui, when he first saw fire, took it into his hands. It scorched him so agonisingly that he plunged into the sea, and this island appeared.

After a week at sea, a tremendous gale arose from the west, and our vessel rose to the heavens before plunging to the depths. We expected at any moment that she would go to pieces, but by the Providence of our God we were preserved and found shelter in Hicks Bay. We found relatively few natives, but those we did encounter were eager for religious tracts. I fear these tribes suffered greatly before missionary efforts stopped the Ngapuhi in their murdering ways.

We made our destination of Port Nicholson and were greeted by a grotesque party of natives, some bedaubed with red-ochre and oil and others with congealed blood. However, they gave us a hearty welcome. Some of our lads, slaves for the last ten or twelve years, met relatives and friends who they greeted with tears of joy and wailing. Port Nicholson is an extensive harbour with plenty of water at the entrance, surrounded by a chain of beautiful hills. There are numerous villages along the shore, and the people there seemed milder and of gentler carriage than those of the north.

On the Sabbath we held a service, with a boat-sail to shelter the preacher. Almost all attended. Mr Hobbs had not proceeded far when the rain came down in torrents, but the people seemed intent on hearing the Gospel and remained listening.

We met with one of the chiefs, Wharepouri, and selected some land for a Mission by way of tapu, in exchange for some blankets and fish-hooks. This should give us the right to purchase at a later time. We also left books, slates and pencils for the lads we left behind, so that they may start to teach these barbarians. The lads were almost broken-hearted when we left, and I trust that they will be as lights shining in a dark place.

The next part of our journey was again uncomfortable, as we were tossed around like a cork for thirty hours, with light winds which meant we were unable to make headway. We arrived in the Southern Island, which looked barren and repulsive, and anchored in Cloudy Bay. Stupendous mountains were covered in snow, and rocks rose directly from the sea. Our native audience there was small, augmented by a few Europeans from the whaling stations. One woman, who I took to be Mrs Guard, was there, but we were not introduced. She looked a striking woman, and I thought of your fascination with her.

We next went by moonlight passage into Queen Charlotte's Sound, a romantic place but with few opportunities for agriculture. Again, the native people here are keen to read and already observe the Sabbath, untainted by the rough European whalers they live beside.

The most notorious person we met with on our journey was Te Rauparaha, who we visited at Mana Island. He has been a great but vicious warrior. He and a number of his men were in a large native house filled with smoke and heat, such that we had to make our excuses and leave earlier than he would have liked. He came aboard the next morning for breakfast and told us that if we would give him a missionary, he would stop fighting and his people would begin to serve God. We presented him with a copy of the New Testament, and left our native man Paul there, one of our most pious and clever lads, as a Teacher.

We paused at Kapiti Island briefly, and sailed past the splendid Mount Egmont, girded with clouds and capped with snow. We stopped a few hours at Captain Cook's Sugarloaf Islands, where the terraces house a few settlements, but *could not land. Nor could we find any shelter along the Taranaki coastline, until we arrived here at Kawhia.*

Our plan now is to walk the remaining 200 miles to the Hokianga, and hence home to you. I hope this letter will find you before I arrive and reassure you that my work has been necessary and useful. We will stay a while with Mr Whiteley, and then will travel to visit Mr Wallis in Whaingaroa.

For now, I bid you adieu, and look forward to the comforts only your dear attentions can supply.

Your loving brother,
John Bumby

I look up from the epistle. So much information, so many adventures. I feel embarrassed that I've been wishing him home when he has been achieving such great things. He is right; I have no desire to share his journey. In every place he stopped, he was never sure of his welcome. And taken on top of that, the torture of the sea voyage.

I pick up the papers. I'll take them over to Jane, to share the news.

I seek indications of John's return every day after that. I make sure I have plenty of food ready if he should arrive, and sometimes need to give it away to my helpers instead of letting it spoil.

It is a long three weeks later that he arrives home, at one in the morning, feeling rather poorly after a rough journey. His neckcloth is filthy and spoiled, he has buttons missing from his jacket, and his hair has grown overlong and curls over his collar. I fuss around him, providing morsels to pique his appetite, and fetching clean warm clothing.

'Fourteen weeks, it's been!' I exclaim. 'Three months away from home comforts. You take too much upon yourself, John.'

'No, indeed, I did experience some wonderful hospitality during my travels,' John says. 'Particularly from Mrs Whiteley, who undertook to darn my socks. I hope you get to meet her. She was especially gracious considering they have recently lost a child.'

That is another thing I can be grateful for; being a spinster means I don't have to live through the agony so many parents face when a child dies. I remember how devastated my own mother was when my sister Jane died. Mother never fully recovered from her loss.

'Well, you can now stay here and concentrate on your local congregation,' I say. 'You should see the progress on the new chapel. And I'll have to introduce you to some of my pupils, so you can hear how much closer to the Lord they've become.'

I have started classes for the girls, despite John's discouragement. Nancy has joined me and is a useful teacher, with her greater knowledge of their language, although I gather that she never learnt her own letters. I am keen to tell John all about it.

'Oh Mary Anna, what a blessing you are to me. But first, I would like to reacquaint myself with my bed.'

The weather the next morning is even worse, with rain lashing down from dark grey clouds which withhold the light. After breakfast, I allow myself the comfort of sitting by the fire, knitting some new socks for John. I've just come to the tricky part of turning the heel when I hear him come from his bedroom. Setting aside my knitting, I look up at him. He is rubbing his eyes and is still very pale.

'Here, sit with me by the fire,' I say. 'There's little else that can be done today. Thank the Lord that you're not out in this. Listen to that rain on the roof!' I am sure that if I look out the window, I'll hardly be able to see the river, but there will be rivulets coursing down the hill towards it.

'Don't fuss, Mary Anna,' John says. 'Later on, I must write my report to the Society. But yes, I will join you by the fire.'

Nancy, alerted by the footsteps on the floorboards, comes into the room. 'It's good to have you home, Mr Bumby sir, it is. I'll bring you a cup of tea. What would you like to be eating?'

'Thank you, Nancy. Tea would be lovely, and maybe a piece of toast.' John rubs his eyes again.

'I received your letter, Brother. It both astounded me with what you had done, and horrified me with your travails,' I say. 'I imagine you were pleased to quit the ship and travel overland.'

'Oh, my letter,' John says. 'I'm happy that it reached you. Did you know that Mr Hobbs received a letter from his wife and children as we were coming home? Mrs Hobbs had addressed it to the Reverend Hobbs, Down South, and the messengers found us. Isn't that astonishing? Remember, Mary Anna, when we were in London nearly a year ago, and we saw that great Post Office? I shouldn't imagine they'd believe a letter addressed thus would find us.'

'Indeed, if I had known it was so easy, I would have sent one myself,' I say. 'Nancy, could you fetch me a cup also, please?'

John stretches out his legs. 'We would have been home sooner, but we were called on to avert a war,' he says.

'A war?' I echo. 'How were you involved? What happened?' I'd been thinking that their journey had been the worst of their experiences, and now this!

'Some of Whiteley's Christian natives warned us that the Kawhia tribe, with their leader Ngatapu, were travelling the country to exact revenge on another tribe for an insult they'd suffered. We were reluctant at first to get involved, but felt it was our Christian duty to do so. It was an arduous task. Some of our natives went ahead with axes to cut a track through the forest. Not only that, but we struck extensive swamps, rugged mountains and deep rivers. It took us a week to get to the pa.'

'Oh John, no wonder you are exhausted.'

'We did see some wonders on the way,' John says. 'In the distance we could see two volcanoes with snow on their summits. It was cold, I tell you. Some mornings we awoke to thick frost on our tents. And we discovered a cave with thick spars from the dripping water.'

'You may have been the first Europeans to see such things,' I suggest.

'I suspect we won't be the last. There was some very rich soil along the way, with fine potatoes growing. Anyway, when we reached the village, we found the two chiefs there, Taonui and Tariki, ready to resist any attack to the death. They were pagans and unwilling to accept our counsel.'

'What then could you do?' I realise that my mouth is still open, focused as I am on hearing John's story.

'A messenger arrived to say that the Kawhia people were approaching. Oh Mary Anna, I will never forget their reactions. They distorted their faces into hideous forms, brandished their spears, uttered the most horrid yells, and appeared more like incarnate fiends than human beings. Then they went to prepare their muskets. It appeared to us that all was lost. But Mr Whiteley offered to go to the advancing army and try to negotiate, while Mr Hobbs and I stayed with the threatened tribe. Finally, it was agreed that, because we missionaries had interfered, they would not proceed to a full battle.'

'Thank the Lord!'

'They still proceeded to stage a war, though. The two sides assembled on either side of a valley with a white handkerchief tied on a pole in the middle, where we stood. On either side of us were hundreds of savages, naked apart from their belts and cartridge boxes, all ready for action at a moment's notice. Then they fired their muskets into the air and did a war dance which shook the ground before dispersing. Honour was thus maintained, on both sides.'

'The lives you saved,' I say. 'The souls of those who saw the power of our belief, who will be more amenable to our preaching. It was great work, John.'

'Yes, I do believe so. If the war had begun, then we were told it might have spread throughout the surrounding districts, from the great lake over to Tauranga on the eastern coast.'

'And then you had to travel back to the coast?'

'Indeed. We stopped along the way at various villages to give services, and finally reached Mr Wallis' excellent Mission Station at Whaingaroa. At the native village nearby was the largest place of worship I have yet seen in this country. We left Mr Whiteley there and began the last leg of our journey through the Manukau and Kaipara districts. I must say my feet were glad when we reached Mr Woon at Pakanae, and he kindly brought us up the river last night by canoe. Mr Hobbs was so happy to be back that he sang a rowing song, waking everyone along the river.'

'I didn't hear a thing until you tapped at the window,' I say. 'I've never been so glad to be woken in the middle of the night. My prayers have been answered. Now, to get you well again.' I rise and put another log on the fire, then sit back with my knitting. When Nancy has had a chance to wash John's travelling clothes I expect there will be plenty of mending to be done, so I'd better finish these socks first.

The next morning, John does not rise for breakfast. Instead, I can hear him calling my name. In alarm, I enter his room.

'Mary Anna, I fear I have gone blind,' he says, rising on one elbow as he hears me come in. 'I cannot open my eyes.'

I look closely. His lashes are gummed shut with yellow crud, and his eyelids are red and inflamed.

'You have an infection,' I say. 'I told you that your health would suffer if you went on such an arduous trip.' I sigh. 'Lie back and I will get something to clean them up.'

I send Nancy down to the wharf to get a bucket of clean salt water, while I find some soft cloth. The water is heated over the fire until it is tepid. Then I take a towel into John's room and place it, folded, under his head.

'I'm going to bathe your eyes,' I tell him. 'It might sting a little, but you must be patient and wait until I'm finished.' Then I gently dab at his eyes, soaking the crusted secretions until they soften and can be wiped away.

'Oh Mary Anna,' John moans while I work. 'How can I do my duties if I cannot see? Not to be able to read, or to write. I won't be able to travel to the other stations.'

'Stop fussing, brother,' I say, continuing my ministrations. 'We will do our best to restore your sight. Hemaima tells me they have a plant which is useful for this condition.'

'I can't have their pagan treatment!' John says.

'Why not, if it works?' I ask. 'They know the trees and bushes around here better than we do. If we were at home, then I would be able to get eyebright or chamomile or marigold, but I haven't seen any of those growing in the gardens. Now, lie still!'

While I am worried, I think it won't do John any harm to be confined to his bed for a few days. I cancel classes to spend the time near him, sewing if he sleeps, and reading to him when he is awake. His eyes continue to ooze until Hemaima brings in the kawakawa poultice. I add some of my meagre supply of honey to it. John forgets to protest, so pleased is he with the relief it brings from the stinging and concern.

'Come, sit by the fire with me,' I say, when he is feeling better.

We sit together companionably. I am content, having him safely back at home.

'I'm not sure that I'm the right person for this life,' John says after a while. 'I know I feel it deep in my soul that I'm called to save these people. But I'm not rugged like Mr Hobbs, who seems to tackle any physical obstacle with relish. My mind is willing, but my body is weak, despite my prayers.'

'Anybody would feel like that, after the journey you've undertaken!' I reply. 'You just have to let yourself recuperate.'

'No, I've been thinking it for a while,' John says. 'And I've been considering the idea that after we've done our duty here for a few years, perhaps we should return home.'

I feel conflicted about the idea of going back to England. If we are to be only temporarily in Hokianga, then I'll treat my life here differently. Recently I've become more settled, believing that we are here indefinitely. I've begun to plan accordingly. I wonder whether it's John's exhaustion and illness making him talk so. He's always been inclined to be morose, with fits of it overtaking him at times. Yet, before he and Mr Hobbs set out, he was energetic and excited about the prospect of new mission stations.

'No need to decide yet, though,' I tell him. 'You may feel differently about it when my summer garden is in bloom, and when I have my own honey again.'

Maybe that is what he needs – some honey to sweeten him up!

Mr Turner has arranged passage on the *Francis Spaight*, bound for Van Diemen's Land. No easy task, since his family is a party of twelve. He has continued to suffer from his lung condition throughout the winter and has only been waiting for Mr Hobbs and John to arrive back before he can abandon the mission.

I have my girls help take the family's possessions down to the wharf. It is surprising how much they've been able to fit into their small native house. I wonder how much more there would have been if they hadn't lost so much in the fire the previous year. All their keepsakes from their years in Tonga especially, like Mr Hobbs' collection of shells that he carefully displays in his house.

I haven't grown especially close to Ann in the six months I've been here, but I am still sad to see her go. The Turners seem to be constantly moving, and I'm not sure whether it is because the Society needs them to, or because Mr Turner is always looking for a better place. The birthplaces of their children show the family's fickle nature. The first children were born in the Whangaroa mission, further north than here, which was abandoned due to fears of raids by the natives. Then they moved to Tonga, with three more children arriving. Next was Hobart Town, for a further three, and then two more children, born here at Mangungu. And now, who knows how many more children are to come, back in Sydney or Hobart Town? I feel very sorry for Ann, and grateful that her eldest daughter is of an age to help her mother.

The three youngest girls are running around the deck of the ship as Ann and Mr Turner say their goodbyes. Mr Hobbs seems the most affected.

'You are returning to the place I first met you,' he says. 'Some of the best times I've had were spent with you, at Whangaroa as well as here. I've watched your children grow and have walked beside you.'

'And if it weren't for you, my friend, my wife and children might not be here today,' Mr Turner says. 'Especially Josiah. It's been a privilege to work with you.'

I am in tears, as are most of the adults. It is so touching to see their friendship. Will I build such strong ties over the years I am here, however many they are?

And then they leave. The mission feels very empty without them, even though the children were generally quiet and well-behaved. The Hobbs children miss their company, particularly Richard, who has no other English boys to play with.

Shortly afterwards, Nancy decides to leave. The servant girl is company on short winter days, someone I can talk to without the difficulty of translation. I fear I am to become even lonelier.

'What do you mean, you're leaving?' I ask. 'Are you going back to England? How can you afford the fare?' I pay good wages, I believe, when you consider Nancy also gets bed and board, and I give her a half-day every Sunday after services. But the fares back home aren't cheap, even in steerage.

'It's just that I've met someone, Miss,' Nancy says. 'A man. And we're going to be married.'

'A man?' I echo. 'What sort of man? Shouldn't we meet him first, to make sure he's suitable? You are our responsibility, after all.'

'We was hoping Mr Bumby would do the honours, Miss. He's a decent man, my Davy, a sawyer at the yards at Horeke. We've been seeing each other for a while now.'

'And he can afford to keep you?' I ask. I hope he is a chapel man.

'He is keen to do so, Miss. To keep us both.'

'Ah,' I say. 'You are with child, then. Nancy, I'm disappointed in you.'

Nancy stares at her boots. 'Yes, Miss. I'm sorry, Miss.'

I'm not convinced. 'Oh well, it's done now. Would you like me to write to your family?'

She nods. She still hasn't learnt her letters, despite our classes, and I imagine at the other end of the letter's journey her family will need to find someone to read my missive.

I wonder how they'll feel, hearing that their daughter and sister is marrying and having a bairn so far from home. I feel as if I've let them down somehow, by not protecting Nancy from the depredations of the rough working men who live nearby. Yet what was I supposed to do? I thought Nancy was exploring the forest tracks for some peace and quiet in her time off, rather than meeting up with some fellow.

'And would you like me to speak to my brother about a service?' I ask. I peer more closely at Nancy's apron. She isn't as far along as Mrs Hobbs, so it won't be overly embarrassing to see her wed.

But she is so young. I know plenty of girls of Nancy's station are married early, but that is back home. How will she cope with a young one and a husband away working long hours? At least she has homemaker skills.

Nancy is bobbing her way backwards to the door, chattering away.

'Thank you, Miss. I'm right sorry, I am. But Davy will be surely pleased to hear the news.'

At least someone is happy, I think.

CHAPTER 11

Spring arrives and numerous trees burst into flower. Most magnificent of all are the dangling yellow blooms of the kowhai, which attract the tūī and korimako in great numbers. My bees have no problem finding them and nip among the birds, climbing into the flower itself to harvest the nectar. And one of the vines growing in amongst the trees unexpectedly bursts into large white flowers that the bees seem to enjoy as well. I am relieved, as I've been worrying about what will become of my bees if there aren't many flowers in this country. I can't keep feeding them with sugar. But it seems the winter dearth is now over.

I visit the hives every few days with Marianne and Heni, watching the bees return with multi-coloured balls of pollen on their legs – orange, red, yellow and pale brown. I get Heni to heft the skeps as I try to decide whether I need to split each hive to provide sufficient space for the growing numbers of bees.

Now I face another problem; if I split the hive, where will I house the new colony?

I have seen how the native women use various plants for weaving, making the knee-length skirts they call patai. The fibres of the harakeke plant are especially long and suited to clothing and rope making. Plenty is now being harvested and sent off to Sydney, proving a lucrative crop for the natives and their European trading partners.

Although I visited the skep-maker back in England, over a year ago, and tried my hand at the craft, I don't think I have the skills to build a skep strong enough to survive the heat and humidity.

I show Mata the hives, using gestures to replace the words we don't share. Mata stands well back – she's been stung before and isn't eager to repeat the experience. The next day she brings an older woman from the village to see me.

'House for biting flies,' she says. 'Auntie make. Good.' Mata mimes weaving with her hands.

'Kapai. Ka mihi,' I say in thanks. I wait while the woman presses her nose against mine, our eyes meeting at a close distance. I know enough about relationships by now to understand that the woman may not be a true aunt but has been given that title as an honorific.

Auntie has a wrinkled face dominated by a chin tattoo, a moko kauae, of bold lines. I am used to the tattoos many of the native people wear, although I can't read the information that is encoded into the patterns.

I show Auntie the hives and bring out the cow horn and bone needle that were used to make them. I mime the process, trying to remember how it worked. I break off a piece of the hardened outer shell of the skep to let the woman see the underlying structure. Auntie nods gravely, then walks away with the tools.

'Ae, she make,' Mata says, as she returns to the kitchen.

Auntie returns a week later, a skep in her arms. I go to find Mata, hoping she will again act as interpreter.

The skep is a near-perfect copy of the original, although the different materials used made it look shiny and silky. Auntie has used reeds of some sort, bound with the vine that is so common in the forest. When I press down on the top, it feels sturdy. I check inside and can't see many pinpricks of light. All it needs now is the cloaming. I'll have to ask Jane for some of their house-cow's leavings.

'Ka mihi. He pai,' I tell Auntie, impressed with the workmanship, and give her a little sack of sugar. Auntie offers a small smile, and holds out her pipe, so I fetch some tobacco as further payment.

'When I have some honey, I'll give you some,' I promise, knowing that she doesn't understand English but thinking she'll recognise my tone of appreciation.

'Haere ra,' the woman says as she turns to leave.

'Wait!' I blurt out. 'Mata, could you ask Auntie to make me three more?' I hold up three fingers to the woman.

'Toru?' Mata confirms. 'Mihi Pumipi, why you want more?'

I don't know how to explain that the bees will increase in numbers, sending out colonies to find their own homes. Already the skeps are getting heavy with nectar and pollen. I need to put another skep on top of the first, so the bees can move up and storefy it. And I need skeps for collecting the sister hives which swarm.

'Marianne,' I tell Mata. 'Marianne will tell. But Auntie bring three more. Toru?' I'll explain it to my young helper, whose language skills are so much better than my own. Still, without Nancy around anymore, I feel as if my own grasp of the language is improving. I am starting to settle in.

One of the packet ships coming from Hobart Town brings mail from England. My heart almost skips a beat when I see packages addressed to me. Nearly a year has passed since we left home, but I am still remembered by friends there. I can hardly wait for John to go to his study so that I can open and devour them.

One is from Emma Sells.

Dearest Mary,

It is not the same without you here. No other person teases me the way you did, and sometimes I think they believe I am completely dim-witted. You, though, could tell I was putting on an act. I do love to gossip, though!

The latest news I have for you is called the Bedchamber Crisis, which sounds scandalous enough. But as you may know (although I know you don't follow the royalty as I do), the bedchamber refers to Her Majesty's private rooms. What happened was that Lord Melbourne threatened to resign as Prime Minister, and our Queen Victoria asked the Duke of Wellington to resume the role. After all, it was only five years ago that he was in charge, and he'd spent seven years in the role, so he knew what to do.

But he refused! Then She asked Mr Peel to form a government. He agreed to do so on the condition that she dismiss some of her Ladies, as they were wives or relatives of his opponents.

Father thought that was rather harsh on the young Queen, but we all thought she would do it. And of course, Mr Peel also had experience in leading the country, although not for as long as the Duke.

But Queen Victoria refused to be rid of her Whig ladies, saying that the Ladies were her close friends, not objects of political bargaining. In the end, Lord Melbourne was persuaded to stay on.

It was a very strange time, not knowing who was going to lead the country, with three candidates for the position. And indeed, not knowing which party would be supreme. I know you would be proud, as I am, at our monarch taking such a strong position against these old men. Though I daren't say that to Father!

How are you faring in your pagan country? I have received letters you sent on the journey. It sounds as if you had a horrific time of it. I'm afraid that if we are ever to meet again, it will have to be in England. I don't plan be go further than Florence. Even Father can't persuade me to join him on a trip to India.

Enclosed with this missive is a small sample of the new Darjeeling tea. It is from India! I like it even better than the Earl Grey. Still a bit of a secret, as the Chinese won't be happy we're growing tea ourselves.

All my dearest love to you,
Emma

I pick up the small parcel and shake it. The sound is of tea leaves in a tin. I hold it to my nose and can faintly smell the musty greenness of the leaves. I hope they've been well sealed and dry for their long journey; I'll be so disappointed if it has gone mouldy before it reached me.

It's unsettling to hear about all the political machinations that have been happening back home. I like to think of England as being strong and stable, but all that uncertainty makes it seem less secure. I take a deep breath. There wasn't anything I could do about politics when I lived there, and now it is all happening at such a remove, it really isn't worth worrying about. I'll leave that to Emma.

The next parcel is from Sophia Hyde and turns out to be a collection of books and magazines. I send blessings to Sophia for remembering.

My very dearest Mary,
I am sure you will enjoy these, as I have. Every time I read something I enjoy, I put it aside to send to you. Our wonderful Boz has finished The Adventures of Oliver Twist, so I send you the full collection. Lucky you, not having to wait a month for each new instalment. I know you read some of the early issues, but I've decided to send all the Bentley's Miscellany so that you can read from the start. He's started a new series, The Life and Adventures of

Nicholas Nickleby, but I will collect those and wait until it is complete. It must be nearly finished, I think.

Among the books is one of poems from Elizabeth Barrett Browning, and Nan Darrell, or the Gypsy Mother. It seems that gypsies are becoming a popular topic of interest.

The chapels in Birmingham continue to be well attended, although none of the preachers has a way with words as your brother does. We still cannot believe that you have gone from our shores, despite me being there to wave you off.

Our greatest news is that we are expecting a little Hyde to join us. At last! Everyone is saying, most especially us. The Lord has finally heard our prayers on this matter. He is to arrive in autumn, according to the physician. William is looking forward to being a proud papa. I remind him that it may be a daughter, but he is sure it will be a son. If only he could be baptised by Mr Bumby. I am now forbidden to go far from home.

Mary, some days I miss you so greatly that I don't know what to do. Then I think, 'What would Mary do?' and I visit the poor, or one of the elderly, and it takes my mind from myself. You are such a selfless person, to have gone with your brother all that way. I do hope you are considering coming home one day.

With fondest wishes from both of us (soon to be three!),
Sophia and William Hyde

I am in tears by the end of the letter. I'm not sure why, as the news is only good. Is it because it brings back to me how much I miss them all? Not only these friends who have written, but all my acquaintances back home. The events I am missing, such as Sophia's confinement. Life is moving on, but I'm not there to see it. And Emma's mention of the horrors of the sea voyage only suggests half the story because I never described to anyone just how bad it had been. I'd love to return to these friends, but the idea of repeating that journey in reverse is chilling. I don't think I can survive it, not for any reason.

While John mentions returning, and my friends beg me to come back, I can sometimes deceive myself that I can do it. But deep down I think it would be impossible. A beautiful dream, never to be realised.

Someone is running across the compound, yelling. As they get closer, I can make out some of what they are calling.

'Mihi Pumipi, Mihi Pumipi,' they are shouting. It is Heni, and he is panicked. He stands at my back door, his chest heaving and his eyes wide, like when they do their war dances. Now that he's found me, it seems he isn't sure what to do next.

'Pi, Mihi Pumipi, pi pi pi.' His fingers wave in the air to show a cloud of bees. A swarm.

'Good, Heni,' I nod. 'Ka pai.' I make a *'calm down'* motion with my own hands, and then with one hand, a *'wait there'* signal. I need my veil and gloves. Although, with a swarm, bees often are much less inclined to sting.

When I return to the kitchen, I send Heni off to find Marianne. It will be good for her to observe this as well. I walk over towards the churchyard, noticing that the other native workers are all sheltering in dark corners. Not many are fond of the biting flies, as they have named the bees.

Over the skeps there is indeed a cloud of bees circling, preparing to leave. I'm pleased that Heni has noticed them early in the process. It will be so much easier to recapture them if we can follow where they go. I wonder from which of the hives this swarm has come. I'd hoped that by adding the new skep on top of the others to create a duplet, I'd given them enough space and staved off swarming. Maybe the new queens were already started before I'd done that.

It is a beautiful spring day. The bees couldn't have chosen a better day to start a new colony. The sun is starting to get some warmth to it again, and there is no breeze. The few young oaks that Mr Hobbs planted in the graveyard down near the river are coming back to life, their leaves unfurling in pretty shades of pink and yellow, while the weeping willows are a lime green. Even the normally forbidding dark of the native forest is showing tender new growth.

Heni returns with Marianne, who's thought to bring her own veil and gloves. I send him away again to get his veil and one of the new skeps that Auntie made. Then I explain what is happening to Marianne.

'This is what nature intended bees to do,' I say. 'Just as the birds build nests and lay eggs in spring, so that new birds are created, then bees will create new bees. When they run out of room, the queen will take half the worker bees and move to a new home.'

'But what happens to the others?' Marianne asks. 'The ones left behind. If the queen is gone, then they will die out.'

'That's where they are so clever,' I say. 'As they are getting ready to leave, the bees start making new queens. The old hive will have a new queen, and the swarm will settle down with the old queen still ruling.'

'The bees make a queen? Surely only a queen can make a queen,' Marianne says. 'Royalty is bred, not made.'

'In humans, that's true. But remember, all the bees are from the same queen's line. In effect, the other bees are like princesses. If the queen dies, some of those princesses may even start laying their own eggs. But only princesses that are brought up properly right from the start can become true queens. They have much larger cradles. I'll have to show you one when we harvest the honey. They look like peanuts, still in the shell.'

Meanwhile, the circling bees are moving in wider arcs. Suddenly, a group branches out and flies towards the forest. The other bees follow, the streak of their black bodies vivid against the blue sky.

'Keep your eyes on them,' I say. 'Come, we must follow.'

Heni catches up with us as we near the edge of the forest.

'I don't think they will go in very far,' I say. 'They'll be looking for a place to rest.'

'Is that why we brought the skep?' Marianne asks.

'Sometimes, if we're very lucky, they will find a skep and move in by themselves. It's more likely if it is a used skep, with the smell of honey and wax still on it. But for now, they are more inclined to find a handy bush or tree to land on.'

As if listening to my advice, the swarm starts condensing around the branch of a tree. As we watch, the bees mass together and start hanging from the branch like a teardrop.

'That's amazing,' whispers Marianne. 'Why are they doing that?'

'They're waiting,' I say. 'The queen will be somewhere in the middle, and they are protecting her, keeping her warm. Meanwhile, some of the bees will go out to find their new home. It might be in a dead tree, or in a small cave, or a hole in a tree that a bird has used. When they find something that might be right, they'll come back here and get the others to follow them.'

I wait while Marianne explains this to Heni.

'Heni asks how long this will take,' Marianne says. 'Do we wait here and then follow them again?'

'It can take a few days,' I say, 'and I'm not planning to spend all my time under this tree. No, what we will do is encourage them to move into our skep instead. Then we can take them back to the churchyard.'

'How do you do that?' Marianne asks.

'Stand back and watch,' I say. 'Heni, bring that skep here.' My hands indicate what I want him to do. He is to stand directly under the teardrop of bees, holding the skep up firmly. Then I select a long dead branch from the forest floor. Carefully I raise it into the air, then suddenly strike the branch holding the bees.

Startled, the bees let go of the branch and fall into the skep as a mass. It is like a watermelon falling from the sky. Heni staggers with the sudden weight.

'Wonderful!' I tell him. 'Ka pai.' I motion for Heni to lower the skep, which he does fearfully. We peer into the basket. Most of the bees are crawling around inside, although some climb up the sides and take flight.

'We'll leave the skep here for now,' I say, helping Heni turn it over slowly and settle it on the ground. 'The bees that are still flying about will find it and come inside, hopefully. Then tonight when they are all here, we'll move it back to the churchyard.'

'That's it? That's all we need to do?' Marianne asks.

'Unless they decide they don't like our skep. Then we will have to try again. But if the queen is in there, then the other bees will start to build comb and make it their new home.'

We stay nearby for a while, watching to see whether the bees will stream out. There is some coming and going, but generally it seems as if most of the bees are happy to remain inside.

It is wonderful teaching beekeeping to new people. It brings back to me the marvels of it all. How bees are so happy to accept a new home, provided there is sufficient space and the right companions. A lesson to be learned indeed.

CHAPTER 12

Increasing numbers of ships have been arriving in the Hokianga lately. Many of them are collecting spars and timber from the sawmills that process the mighty trees, growing thickset on the surrounding hills.

When the *Tory* arrives, John invites the Master, Captain Chaffers, to a meal at the Mission House. He brings with him Colonel Wakefield and the Colonel's nephew, Mr Jerningham Wakefield. My girls and I are kept busy preparing the meal, but I manage to join the conversation at the table.

John is involved in a discussion with the Colonel and his nephew.

'And what brings you to this river?' John asks.

'Your neighbour, Lieutenant McDonnell,' the Colonel says.

'I don't think he's there,' I say, 'although you might find his agent, Mr Marriner.'

'We met him in England before we left,' interposes Jerningham. 'He sold us much of his land.'

'He did indeed,' the Colonel agrees. 'Although now that we come to look at it, there is very little suitable for a settlement here. Not enough land flat enough for agriculture. I will claim Herd's Point, near here, then I think we will venture down to look more closely at the land in the Kaipara, which he also sold us.'

'Can I assist?' John asks. 'I think you should consult with the original landowners, to ensure that McDonnell's purchase is valid.'

'That would be very kind of you,' the Colonel says. 'I hope to invite the Ngāpuhi chiefs to McDonnell's Station, but someone who can speak the language would be most valuable.'

'I would be honoured to help in that way,' John says.

I worry that his translating skills aren't up to the task,

but I know that John is keen to ensure the New Zealanders aren't taken advantage of.

'It's a waste, all this fertile land not being used,' Colonel Wakefield is saying, his long face looking serious. 'The natives don't value it sufficiently, but there are many back home who would clear the land. That's what our company aims to do – bring settlers and labourers who can turn it into something productive.'

'I haven't been here long,' John replies, 'but I don't think you appreciate how the New Zealanders work as a tribe. Their land isn't owned individually, and they take a long view of its worth.'

'They still have chiefs, don't they?' the Colonel says. 'That's who we're dealing with. We've had some success already down at Port Nicholson.' He brings out a large cigar and prepares to light it.

'I know you're consulting with the chiefs, but I'm concerned about the settlers,' John argues. 'I fear they won't treat the natives with sufficient respect. The New Zealanders haven't experienced much in the way of commerce, so it would be easy to take advantage of them.'

The Colonel scowls as he takes his first puff. 'They seem savvy enough to me. They know how much we value land. For example, I know you believe that you'd reserved some land at Te Aro for a mission station. However, the natives there don't accept that you have any rights to it, like your fellow claims. Don't worry though, we'll make sure you have sufficient space for a chapel and mission house.'

John frowns. 'That's disappointing news. I thought they'd put a tapu on it, to reserve it for us. Yes, I hope that we can indeed work together for the betterment of the people.'

The discussion looks as if it's getting serious, so I seek to divert it.

'And you?' I ask young Mr Wakefield. 'What is your role with the New Zealand Company?'

'My father is the leader of our enterprise, back in England. He graciously allowed me to come with my uncle, as clerk and secretary. I intend to record it all for a book which will serve posterity,' the eager-faced young man replies.

'And your mother? What does she think of you coming to the other side of the world?' He also has the long face of his uncle, with the hopeful start of a moustache emerging on his upper lip.

'Sadly, I never knew my mother, as she died within days of my birth,' Jerningham says. 'We were brought up by my Aunt Catherine, one of my father's older sisters.'

'We?' I ask. 'You have other brothers, then?'

'No, I am my father's only son, and now his only child. My sister Nina died in Lisbon four years ago.'

'Oh, I am most sorry to hear that,' I say. 'You must miss her. I had a sister who died of consumption when I was about your age.'

'Nina also died of consumption,' Jerningham says, and we exchange sad smiles.

John has been talking to Captain Chaffers, who has bushy sideburns and a fierce glare. He has a forceful manner, and I think that the Wakefields seem a little scared of him.

'She's a good little ship,' the captain is saying. 'Set a record for our voyage out here, would you believe? Only 96 days, from Plymouth to landfall in New Zealand.'

'That's astonishing!' John says. 'Where did you arrive?'

'We were heading for Port Nicholson,' Captain Chaffers says, 'but adverse winds meant we were driven into Ship Cove. There, we took on provisions, anchoring in near the same position as Captain Cook.'

'Is this your first time in New Zealand, Captain?'

'Not at all. I came into the Bay of Islands with the *Beagle*, back in '35. I've been interested in returning, though. I might stay a while in Port Nicholson, as harbourmaster for the Wakefields. There's certainly plenty of coastline to explore. We've named the channel into Queen Charlotte Sound, after the ship. Tory Channel.'

'And you, Colonel? What is your history? Did you get your commission in the Burma War, perhaps?' I ask the older Wakefield.

The Colonel adjusts his necktie. 'Ah, no, I fought in the recent Spanish War. While my oldest brother and young Jerningham here were in Canada.'

'You are an adventurous family,' I put in.

'Another brother of mine is coming out to New Zealand soon,' the Colonel says, puffing on his cigar. 'Arthur is to head up a new settlement, just to the west of the Sounds. We're thinking to honour our great military leaders, Horatio Nelson and the Duke of Wellington, when we name these places. We have great plans indeed. The name of Wakefield will be well-known to history, I can tell you that.'

John calls a District meeting so that he can discuss appointments in the new areas. He is excited to have all his fellow missionaries arrive. He always does best when he is with his confederates.

I wish that their wives could attend as well, but most of them have children to care for and cannot make the long voyage. Jane is heavy with child now, and not able to move about much. And Ann is gone, although she was never much company. I am kept busy catering for this group of men. I'm always on the fringe, never part of the conversation. Mr Woon comes up the river from Pakanae, Mr Whiteley travels from Kawhia along with Mr Wallis from somewhere near him, Mr Ironside and Mr Creed from Pakanae. And, of course, Mr Hobbs, ever-present and controlling.

I am surprised when I hear the development outcome. Surprised, and dismayed.

'Pakanae?' I say to John. 'You're sending the Hobbs family to Pakanae? But that's miles away.'

John looks away from me. 'Mr Hobbs has nearly finished building the new house at Waima, for when Mr Warren and his wife come out from Van Diemen's Land. So, we feel he is ready for the challenge of Pakanae. We've been having difficulties getting many converts there.'

'But Mr Woon and his family are already there,' I point out.

'We feel it would be better for Mr Woon to move up here to Mangungu,' John replies. 'He will be able to devote more of his time to printing. And when the Conference sends us new recruits next year, we expect that one of them will assist Mr Woon with the bookbinding.'

I'm not happy with the decision, but powerless to change it. I go to find Jane.

'How can you accept this?' I ask. 'Your children are settled here. And you'll have to move with a new bairn. Surely your husband can ask for you to remain here?'

'Mary, I will miss you. You always think things can be altered. But we are mere women, and we must do what our betters decide. It's not the first time I've changed residence, and it won't be the last. Just before you arrived here, we thought we were being sent to Launceston.'

'In Van Diemen's Land? I'm fortunate that didn't happen. But can't you try to protest? Why are you so accepting? I can't stand to think of you living so far away. Who will I talk to now? And if you take Emma and Marianne, then my little helpers will be gone as well.'

'Yes, Marianne has been enjoying her employment with the bees. Never mind. This must be the Lord's work, and there will be a purpose to it. Learn to embrace it, Mary, instead of fighting it.'

I am further distressed when I come across John in the kitchen, a plate of honey in his hand.

'Where did you get that?' I ask him. He has never ventured close to my bees before.

He looks embarrassed to be caught out. 'I wanted to serve some honey to my fellow ministers,' he says.

I peer closer at the plate. 'That's maiden honey,' I say. 'Taken from the comb before the bees have a chance to cap the cells with wax. I hope you haven't damaged much of the comb.' I pause. 'Well, you'd better offer it to them, because it won't last long.' Maiden honey doesn't store well, I've found in the past. Fortunately there are plenty of people around to eat it.

When John decides to explore further north, in the company of Mr Whiteley, Mr Ironside and Mr Creed, I am glad. I'm finding it difficult to forgive him for sending my only friend away. I do worry when the weather turns and torrents of rain swell the rivers, but I know he is in good company.

Jane's bairn has not yet been born when her family make the move to Pakanae. It all happens so quickly. I watch as their boat, loaded with possessions and children, leaves the jetty and heads west. Jane's eyes are swollen, but she smiles and waves. My only consolation is that Jane will be near Mrs Eliza White, who has been a good friend to her in the past. After all, if the bairn was born here in Mangungu, I'd be useless to assist. I've never attended a birthing and have no idea what is needed.

After eight days away John arrives home safely. To a chilly reception. He pretends not to notice.

'Yes, we are dirty and wet and weary, but we had an educative journey. It is distressing to see so many ruined fortifications and desolated villages along the way. I fear that the population is less numerous than it was, despite the richness of the soil.'

'Is that the effect we are having on the natives, then? Are we affecting their health?'

'We are here for their souls, Mary Anna. Their everlasting souls. I don't know what has caused them to abandon their homes. It may be the muskets, or the measles. Either way, it has not been us who introduced those to the New Zealanders, but the people of commerce. The whalers, timber merchants, flax buyers. We have to keep working for their souls. That is why we are here.'

'I hope that introducing them to the Lord will balance out the damage, then. It might have been better if this land had never been discovered,' I mutter.

He looks up from unlacing his boots. 'How can you say that? And it is pointless even thinking that, because it's already done. We must go ahead from here.'

'Always a compromise,' I say, as I take his wet boots out to the kitchen to dry.

Without Marianne and Emma to translate, and Nancy not around either, I am learning to speak the native language much faster. Hemaima and Mata are my usual companions, since Mrs Woon doesn't seem inclined to be very friendly.

John isn't back home for long. The other missionaries return to their stations while he leaves on the *Melrose*, bound for Sydney. From there he intends to also visit Launceston and Hobart Town, to see Mr Waterhouse and get further instructions for his work. With him, he takes my letter to our father. So, when Christmas arrives ten days later, I am on my own.

I find myself pausing in the middle of tasks, reflecting on all that has happened during the previous year. It feels so strange to have Christmas without the snow and cold. The first of the plums are ready, despite the ravages of the kererū, and many of the vegetables I've planted are nearly ready for harvest.

I find the forest a pleasant place to pass the time. It is cooler under the shade of the massive trees, and the dappled light means that I can take off my bonnet and gloves without fear of becoming sunburnt.

As Mr Hobbs mentioned soon after I arrived, one of the local trees is a glorious sight in full blossom. The noise of my bees draws me to them. Strange bright red threads tipped with gold hang in bunches, and the bees dance over them, getting covered in pollen. This must be the pōhutukawa, the New Zealand signal of Christmas.

Back at the Mission House, I seek out Heni. He is now helping me look after six skeps, as we've caught three more swarms. The churchyard is filling with hives. One I named Marianne just before she departed. The most recent is called Edward for the new Hobbs bairn, and the third is named Richard. The young boy was delighted to have his own hive, although it didn't encourage him to come any closer.

'Pōhutukawa,' I tell Heni. 'Pi, pi, pi,' and I indicate with a finger that the bees are busy in the trees.

Heni wiggles his fingers in acknowledgement.

Then I mimic lifting the hives, and Heni tells me that they are heavy.

'Time to harvest some of the honey, then,' I tell him.

We fetch bowls and cloths from the kitchen. Hemaima and Mata come with us a short way, curious to know why I want these things, but when they see it is to do with the bees they retreat.

I love my bees too much to kill them off, purely to steal their harvest. I know many apiarists use brimstone to destroy the bees and allow free access to all the honeycomb. I prefer to sling them, to smoke them into calmness and only take what I can. Mr Webster felt the same way.

I start with the hive I've named Sophia, which has the duplet skep on top. I get Heni to lift the duplet from the bottom basket. It is difficult, sealed on with propolis and heavy with honey. We shake the bees from it and turn it topsy-turvy, being careful to keep the comb vertical so that it doesn't break unexpectedly. Wedges of honeycomb hang from the spleets, golden in colour with sparkling white wax cappings. A few bees remain, crawling over the surfaces. I reach in with my gloved hand and break off a piece, my fingers squeezing honey from the comb where I hold on. I drop the broken comb into a bowl and cover it with a cloth. Honey drips from the fragmented edge, bringing the aroma of glorious sweetness to the air. Heni sniffs, then smiles, delighted with the smell despite not having tasted honey before. The circling bees also notice and started to congregate on the upturned skep. Quickly, I break off further pieces of comb and place them in covered bowls before indicating to Heni that he can replace the duplet on the skep.

By the time we have inspected and claimed honey from the other hives, the bees are very excited. Some of them are working to discover entrances through the cloth to our bounty. Others are clustered outside the hive entrances or buzzing around the skeps, smelling the honey, unsure what has happened.

Heni and I carry the bowls into the shade outside the kitchen. Some of the bees fly away, not enjoying the dim light and being taken so far from their home. I flick the few remaining bees from the cloth and spread a blanket over the bowls.

Inside the kitchen, I assemble clean jars and crocks with lids before instructing the girls to leave. Heni is reluctant to come into the kitchen, but I insist. I wonder, does he believe it is solely women's territory? We bring the bowls inside and with the door firmly closed I show him how to tip the honey and some comb into each jar, fishing out any dead bees, and squeezing honey from the wax. Our fingers drip with the stickiness. On the table, little pools of honey begin to form, glistening like thick oil in the dim light.

My old mentor, Mr Webster, used to have a honey press that we'd use. The comb would be placed in the barrel, then a screw would be turned to lower the lid, pressing all the honey out through a funnel. We would be left with the wax, which could be washed and ready for use. However, the press was cast iron and very heavy, and I decided not to bring one with me when we came out. It would be more efficient, but our fingers can be quite effective.

In one of the bigger bowls, I carefully collect the compressed wax. Later I'll melt that down for candles, dipping the wick in repeatedly, to build up layers. John will be pleased to have wax tapers for services. Some of the wax will be used for sealing jars of jam, and more can be used for polishing furniture and floors. It is valuable in so many ways, I can't imagine how people live without it.

We scrape the last of the honey from the bowls with fingers and spoons and seal the jars. Every surface now is tacky. Only then do I show Heni that it is alright to lick his fingers.

His eyes widen. Then his brows come together as he tries to reconcile the taste with his idea of bees. His face erupts into a huge grin, then his tongue emerges to lick all over his hands. He giggles as he eats. 'Pi,' he says with a chuckle. 'Pi.'

'No,' I say. 'Honey. Honey.'

'Heni?'

I point at the jars. 'Honi,' I say.

'Honi tino pai,' Heni says.

I take that to mean it is very good. Heni will now have a greater appreciation of what he is doing, working with the bees. I expect that when the others of his tribe try honey it will increase his mana, and I'll also have a few more volunteers for the work.

Although the summer solstice has passed, the days become even hotter and my clothes cling to my damp skin. I schedule my activities to be outside during the cooler early mornings, avoiding the kitchen whenever possible.

Bad news arrives. Not about John who, as far as I know, is enjoying the company of Mr Waterhouse and other brethren in New South Wales. This is news of a fellow missionary, although not someone I've met.

The Reverend John Williams was working for the London Missionary Society in the Pacific Islands. He was well-loved in places such as Samoa and Rarotonga. But he visited an island further north, where the people did not know him. There, they murdered him and his companion and then ate them.

This is one of my worst nightmares, *cannibalism*. It seems incredible to me that the peaceful, loving people of Nene's tribe ate their foes, and not so very long ago. In some ways, it shows the good that my fellow missionaries have achieved by teaching these pagans the abhorrence of the practice. Still, I worry that it is still happening in isolated areas of New Zealand. So, John's exploration of the country might bring him to the same fate as poor Reverend Williams.

Then one evening, after John has been gone for over six weeks, Mr Hobbs visits. He has a letter in hand and is puffing from exertion.

'Miss Bumby! How pleased I am to see you.'

'Hello. Is everything well? Mrs Hobbs is in good health?' The time around a bairn's arrival is particularly dangerous for the mother.

'Thank you, yes. But I come bearing news of great importance!'

Mr Hobbs looks puffed out like a kererū pigeon, all chest.

'Have a seat, Mr Hobbs. Would you like a cup of tea?'

He flaps his hand around. 'This is not a time for sitting around. I have received a letter from Mr Ironside. He is with Nene and Mr Warren at Waitangi, in the Bay of Islands. He tells me that Captain Hobson intends coming here, with seven other gentlemen. There is to be a gathering of native chiefs, and we are to host them here at the Mission House.'

I decide I need to sit, regardless of my guest. What a time for John to be away!

'What is the occasion?' I ask. 'Where are they coming from, and when do we expect them?'

'Well, you remember how the chiefs met Mr James Busby at Waitangi nearly six years ago to choose a flag? He's the British Resident.'

'Well, I wasn't living here then, but I have heard about the United Tribes flag and seen it flying. It looks very fine to me.'

'And then the next year, Mr Busby got the chiefs to sign the Declaration of Independence, what they call the He Whakaputanga.'

'John mentioned that, when I told him about meeting Baron de Thierry. He says that the Baron's sovereign state claim had worried our people and forced their hand. But that doesn't tell me why Captain Hobson is coming here.'

'Captain Hobson is now to become Governor Hobson of New Zealand. He has gathered another meeting of chiefs at Waitangi and is hopeful that they will sign a treaty with the British Government. After that, they will travel across from the Bay of Islands and hold a second meeting here.'

'That must be why Nene and his men have gone to Waitangi,' I say. 'Most of the important chiefs will have gone there, I imagine. How many do we expect to come here?'

'Well, it is an important occasion. Nene will have sent word out for everyone from the northern tribes to attend, and his people will be preparing to cater for them all. Most are relatives anyway, from what I can gather. But we will need to accommodate our new Governor and his entourage. This letter mentions that a couple of missionaries will accompany him.'

'Maybe it is as well that Mr Bumby is away,' I say. 'I'll give the governor his bed. I might have to borrow some bedding from your good wife for the others, and they'll need to sleep on the floor. Fortunately, it's summer, and still warm.' My mind is already preparing lists and checking supplies.

'Are you certain you can manage, Miss Bumby? I must say, I would rather your brother was here.'

'I am perfectly capable, thank you Mr Hobbs. I run this house, not John.'

'I suppose that is true.' He pauses. 'I just meant that I would prefer your brother to be the host. I am not so comfortable in the limelight.'

'I'm sure you will acquit yourself adequately,' I say, unwilling to concede much. How dare he ask whether I can manage? How hard can it be?

This is my role, after all. I am my brother's housekeeper, here to entertain his guests, whether he is present or not.

Miss Bumby's Mission

CHAPTER 13

It is as well that Mr Hobbs has given me some warning. Less than a week later, Governor Hobson arrives on horseback, along with Captain Nias from HMS *Herald*, Willoughby Shortland, Felton Mathew and the CMS missionaries Richard Taylor and George Clarke.

They aren't the only arrivals. Streams of natives appear, finding areas around the Mission House to set up camp. Native boats are lined up on the shore, and campfires soon produce a haze in the air as the smell of pork and kumara and Indian corn fills every corner.

'How will I feed them all?' I ask, as I watch the New Zealanders settling in.

'Nene,' Hemaima says. 'At our kāinga, we are making a hāngī.'

I have tried native feast food before. The smokiness takes some getting used to, but the pork is tender and juicy, and the starchy foods – potatoes, pumpkin, kumara and corn – help to soak up the fat.

I don't think I can serve native food to our English guests, though. Boiled vegetables and roasted meat, followed by a pudding, will be more to their tastes.

'Miss Bumby, you have done an admirable job in looking after us,' Governor Hobson says as we walk around the mission station after dinner. Mr Taylor and Mr Clarke have expressed a wish to visit the little chapel and the nearly completed new one.

'Thank you, Governor Hobson. I'm sorry that my brother isn't here to meet you. But I'm sure he'll be pleased to know that you have the interests of the New Zealanders at heart.'

We've come to the enclosure where I keep my skeps. The warm day means the bees are busy out collecting, and the buzz of their wings is audible and pleasing. I see the Governor startle as he notices.

'But what do you have here? Are they truly honey bees?' he asks.

'Indeed. Did you not notice the honey that was in the pudding? That was produced from these hives.'

'Where did you get them? Mrs Hobson and I have been asking about bees ever since we arrived and have been told that there were none here.'

'I brought two hives with me from home,' I say. 'I wasn't sure whether they would settle, but they appear to greatly enjoy this climate.'

Mr Taylor joins in. 'When I travelled out, one of my fellow missionaries brought bees with him as well.'

'Who was that?' I ask. 'I haven't heard of anyone else who keeps bees here. I would be keen to exchange notes on their management.'

'Mr Yate. It was his intention to bring them to New Zealand, but I'm afraid he only made it as far as Sydney and was then sent home,' Mr Taylor says. His face goes red. 'Besides, the bees all died before we made it to land.'

I know it is sinful to be so full of pride, but I can't help myself. It took work and determination to keep these bees alive for the whole of that journey. I like the idea that I might be the first to achieve it.

The next day, some chairs and a table are set up on my veranda. At ten o'clock the bell of the chapel is rung to begin the meeting. The Governor addresses the gathered people. He is newly arrived in New Zealand and does not speak the language, so Mr Hobbs obliges by interpreting. Nene has already signed the treaty at Waitangi six days earlier, after speaking forcefully in favour of it. He encourages the chiefs to follow his lead.

I watch as the groups confer. Some seem to think that the English are thieves, and that the Treaty is meant to deceive them. Mr Hobbs repeats Governor Hobson's assurances that the Queen does not want the land, and that it will never be forcibly taken. Finally, at six in the evening, they start lining up to sign the parchment document, which is a copy of the original. The missionaries' work in literacy is evident, as some of the principal chiefs can sign their names, while others just leave their mark.

The next day is one of celebration. It begins with a haka consisting of over a thousand men, which I watch from my window. The glass panes rattle to the thump of the men's feet as they stamp in the dirt. The Governor has gone on board one of the ships anchored just offshore. Then the main hāngī is opened and the feasting begins. I am relieved to see that there is no alcohol, but tobacco is being distributed, and gifts of blankets are given to the rangatira, the chiefs. Even more people seem to have arrived. From nearby Hōreke, Lieutenant McDonnell's cannon fire a salute, answered by those of Mr Russell.

Governor Hobson is very pleased when my dinner is served. He's found it tiring to be under such scrutiny over the last days, with all the unfamiliar names and protocols to learn. He retires to bed early and tells the remainder of the company that he intends the next day to be one of rest.

The following morning, however, he expresses an interest in meeting Baron de Thierry. I'm able to give him instructions on how to reach his house and send a native worker with him. Mr Taylor goes with them. I wonder whether it's Governor Hobson's intention to subtly rub the Baron's nose in the British acquisition of New Zealand, but I don't doubt that he will be charmed by the Baron. They return sooner than expected, prevented from getting there by a dangerous muddy creek crossing.

Then the company departs.

I am alone again, and the world carries on without me. It has been a brief episode of activity which has taken my mind off my distant friends and provided plenty of tasks to accomplish. When I venture to the village, I see that they are also exhausted from the effort of catering for so many. It has taken so few days that I wonder whether it will have much impact on our lives.

I don't have the house to myself for long, though. Only a few days after the Governor's party leaves, John arrives home from Sydney. With him is Mr Orton, whom we met in Hobart Town.

'We expect the *Triton* to arrive at any time now,' Mr Orton explains. I don't understand the relevance of his comment.

'The ship purchased by the Mission Society?' I ask.

'Indeed. Purchased for three thousand pounds, so I've heard,' says Mr Orton.

'Yes, we've heard from other ships that the *Triton* was due to leave Bristol in mid-September,' John says.

'If she makes the same time as us, she should indeed arrive this month,' I say.

'We are expecting reinforcements to our ranks,' John says. 'The Society has listened to our pleas and are sending more missionaries, for which I'm very glad. Having more people will make our work here much easier.'

'When she arrives, I intend to venture to the Friendly Isles along with those others who are heading there,' Mr Orton tells me.

Now I understand. He is using our place as a staging point.

However, the end of summer arrives, and it's the first anniversary of John's and my arrival at Mangungu, with still no word of the *Triton*. Every day we look out for her, fearing the worst.

Meanwhile, I am busy collecting fruit and preparing it for winter. As the days cool, the worker bees start to evict the drones from the hives.

It is hard to explain this to Heni, who is distressed to see the workers dragging their fellows out of the entrance and forbidding them to return. I wish that Marianne was still nearby to help me explain the concept.

'Should I tell him that the drones are slaves?' I ask John.

'Are they slaves? I thought you said they were the males.'

'They are. And they're the opposite of slaves, as far as I can tell. I never see them collecting from flowers. They don't seem to do any work in the hive.'

'Maybe you can tell him that they are rangatira – chiefs,' John says. 'And the workers are the slaves.'

'But that doesn't explain why they are sent outside the hive to die in winter,' I say.

The other thing I ask Heni to watch for is robbing. As resources get low, sometimes a strong hive attacks a weaker one and takes their honey. This is an easier concept to tell Heni, who has been brought up in a world where an attack from your enemy can be expected.

I hope that I have left enough honey in each hive to last the bees through the dearth of flowers. I remember the previous year, when I had to feed them sugar water to keep the hives going. But if I leave too much honey on the hives, it encourages robbing and keeps the queen laying for longer, rather than shrinking hive numbers down. It is a tricky balance.

The equinoctial winds arrive, causing rough seas. And still *Triton* fails to show.

Then at last, after three months of looking, we see Mr Waterhouse coming up the river. I am overjoyed. It has been over a year since I've last seen him. It's almost as if I have my own father back with me again.

'We had thought that the bar would be the worst of our problems,' he says, accepting a cup of tea. He is explaining the difficulties *Triton* had in approaching Hokianga. 'We were giving thanks to the Lord for our safe deliverance into deep water inside the entrance. But then the wind freshened in our faces, and the anchor started to drag. We were in danger of being blown onto the rocks on the north side.'

I remember seeing those rocks. There would have been no hope of saving the ship if it had foundered.

'Captain Beatty put the boats in again, trying to tow us to a safer mooring. Against his advice, I took one of the boats and rowed ashore to find the pilot. Instead, we met Mr Hobbs and some other men, who organised more rowboats to effect a rescue. Eventually we were able to anchor around the corner.'

'The Lord does indeed look after His own,' John says, visibly shaken by the close call of his friend and mentor.

Mr Waterhouse stays with us overnight, and the next day Mr and Mrs Hobbs arrive.

'Jane!' I exclaim. 'How I have longed to see you. Let me look at that bairn.'

The baby is gorgeous, with plump red cheeks and smiles for everyone. Jane and I retire to the kitchen to show him off to Mata and Hemaima, and where Jane can tell me more about the antics of the other children. I have missed their company.

We spend a pleasurable day together, while the men talk. That evening *Triton* brings the remainder of the missionaries to Mangungu. It is a party of twenty that I must care for, but by now I am well used to such numbers. The bounty of my autumn garden means plenty of fresh vegetables can be served, along with the ubiquitous pork.

Some of the missionaries are bound for the Pacific Islands. I spend time with Mrs Buddle and Mrs Turton, who are to stay in New Zealand. The experience of my year in this country is passed on, much as Jane helped me. It is interesting to see those things I have become accustomed to through the fresh eyes of these ladies.

It's a busy household with so many visitors. I'm occupied with supplying meals and keeping the house warm, while the men talk of great deeds and ambitious plans for their ministries. John gives up his bed for Mr Waterhouse and sleeps in a storeroom, which leads to him catching a cold.

Among the new arrivals are three single men: Mr Buttle, Mr Aldred, and Mr Gideon Smales. They are going to be more difficult to place in mission stations. It is felt that married men are better equipped to live in areas isolated from other Englishmen. It is well known that unmarried merchants and traders often take a native wife, of which the Society disapproves.

John is talking to Mr Smales at dinner later in the week.

'I understand that you've been sent to help with the bookbinding, Mr Smales. We'll be glad of the assistance, especially Mr Woon. He's been slowed in his printing by the need to assemble the books.'

'Yes, about that, Mr Bumby. I'd really prefer to start proper missionary duties as soon as possible. To get out among the natives and tell them of the Lord.'

'I'm not sure about that. Circuit work is not where our greatest need is.'

'I've done all the necessary training, Mr Bumby. And I would like to be of use.'

'I understand, Mr Smales. I too have been young and eager to spread the word of our Lord,' John says. 'However, you will have to wait for me to return before I can assess your abilities. Mr Hobbs and Mr Woon are expecting you to assist with preparing the tracts.'

'I'm afraid they may find I am more of an outdoors man.'

'You must pay attention to Mr Hobbs. He is the most fluent of us in the native language, and it is important that we can converse with them. He will also help you learn about their customs.'

'Very well, Mr Bumby. I am your humble servant and will do as you advise.'

I've been listening to their conversation. The mention of John leaving worries me. He is determined to sail with *Triton* down to Mr Whiteley's at Kawhia, along with Mr Waterhouse and the missionaries who are bound for the Friendly Islands and Fijee. Then he plans to travel overland again. I've tried to dissuade him, to no avail. His nose is red raw from his cold and he is thinner than ever with all this constant movement. Whereas, a year of staying in one place has really plumped me out.

'Mr Smales,' I say, hoping to divert the men from their seriousness, 'do I detect a Yorkshire accent?'

'Such as your own, Miss Bumby? Yes, I hail from Whitby.'

'I've never been there, but I hear it is lovely.'

'Famous for its ship building and the training ground of the renowned Captain James Cook,' Mr Smales agrees.

He is quite young, with dark hair and deep-set eyes that gaze fixedly at me.

'And what do your parents think of you coming all the way to the other side of the world?' I ask him.

'My father was uncertain about me becoming a man of the cloth at all, but there was no place for me in his business,' Mr Smales replies. 'And my mother didn't want me to leave, but she has my two brothers still there to console her.'

'And why did you come?'

'That's a good question. Adventure, certainly. Opportunity perhaps. And to spread the word of the Lord, of course. What about you?'

I blush. Few people ask why I came.

'To support my brother in his good work. I promised our mother on her deathbed that I would take care of him. I didn't realise at the time that it would involve such a journey! But we have been here for over one year, and I find it is usually quite congenial.'

'Yes, your brother is a remarkable man. I can imagine following him to the ends of the Earth,' Mr Smales says. But instead of looking at John when he says this, he continues staring at me. I find myself shifting under his gaze.

'Excuse me, I must see what the girls have done to the pudding,' I say, as I rise from the table.

Mr Waterhouse has been kept occupied conferring with each of the brethren, as well as meeting others of the European community and some of the early native converts. It is a joyous occasion to have so many gathered together.

On the Sabbath there is a special service on account of Mr Waterhouse's presence. Nene has sent the word out, and the mission is crowded. Mr Waterhouse baptises ninety in the morning. I am especially pleased with the choice of hymns, which include *And Can It Be*, with its beautiful swooping refrain of '*Amazing love, how can it be! that Thou, my God, shouldst die for me!*'

In the afternoon some of the converts give their testimony, including Nene. Then, to the surprise of many, John leads a hymn in the native language, the first time he has done so. His voice shakes with emotion and gradually the crowd joins in to wonderful effect.

While we are all gathered at Mangungu, we arrange to send a petition to Governor Hobson requesting a postal service to be set up between Hokianga and the Bay of Islands. There is a feeling that we are on the edge of a great upwelling of conversion and civilising of the natives, now that our numbers have increased.

Then it is time for Mr Waterhouse and the Island missionaries to board *Triton* and travel onwards. Some of the native men from the mission are sailing with them. A Tongan man called Joel, who came to New Zealand with Mr Hobbs, is also on board, planning to return with *Triton* when it leaves for the Islands. I go on board to inspect the vessel and find it very small, much smaller than our *James*. No wonder they had such a rough voyage from England.

'Are you sure that you should go with them, John? Surely Mr Waterhouse can stand in your stead? You are not well; I think you should stay and rest.' I feel that the bustle of so many visitors has worn him out.

'It is my duty, Mary Anna. The care of these people lies with me. I will only go as far as Kawhia, and then travel back overland. I want to visit the Thames area and return through Whangarei.'

'But then you will have to return to Kawhia in a few months for the next district meeting!'

'Enough, Mary Anna. I am grateful for your concern, but remember that it is the Lord's work we are doing. What does it matter if I am inconvenienced, when we can bring so much hope to these people?'

Triton moves down the river, although given the weather, it is uncertain when it will be able to get out of the Heads. I stand on the jetty waving them farewell. I can see the tall figure of Mr Waterhouse and the thin form of my brother as they stand on deck, looking back at the Mission House. Mr Creed and Mr Smales have gone with them and will leave the ship before she departs. In the meantime, the mission is back to its silent, echoey self. The wind whistles through the oaks and willows of the cemetery and the clouds scud across the sky. Slowly, I trudge up the hill to the house. There is plenty to do to put my place to rights after all the visitors, but also a great deal of time in which to do it.

CHAPTER 14

I continue to worry about John, despite knowing that he is with Mr Waterhouse and the others. It is quiet with only a few people left around Mangungu. Mr Smales conducts the services on Sundays and works with Mr Woon during the week. I wish there were more visitors to keep me occupied and stop my mind from fretting. It is winter and dismal with wind and rain, which affects me even more. I can picture my brother, caught out in the weather. He is often in my prayers.

Then, quite unexpectedly, I receive a note from Mr Smales.

> *My dearest Miss Bumby,*
> *I know we have not had a long acquaintance, but I wish you to know of my sincerest regard for you. To this mind, I wonder whether I could offer you my hand and my heart. Together, I believe that we could achieve great things in this raw new country.*
> *I would have preferred to discuss this with your brother, but as he is likely to be absent for some time yet, I felt I could wait no longer. I do hope you will consider this and confer the great pleasure of your acceptance.*
> *With warmest wishes,*
> *Your servant,*
> *Gideon Smales*

I am stunned. He has only been at Mangungu for three weeks. I hardly know him with such a short acquaintance, and I'm not looking to marry anyway. He is younger than me, only 23 to my nearly 30, and seems quite naïve. If he wants a wife, why did he not arrange one before he left England?

As I think more about the proposal, I begin to be offended. I am the only woman of marriageable age amongst the Wesleyans. It feels as if Mr Smales has decided he would receive better opportunities as a married man. It is not any of my own attributes that has brought about the offer.

Does Mr Smales not appreciate how devoted I am to my brother? He is handsome enough, with a vitality about him which appears healthy compared to John's wan complexion. Also, the sermons I have heard show that he is well-educated and pious. And he is also from Yorkshire. But that is not sufficient reason to abandon John.

It feels awkward to have him living so close by. I might have to watch what I say when we meet. And I imagine that I'll be searching his sermons in case they have some other meaning for me to interpret.

Dear Mr Smales,
Thank you for your kind offer. However, I am not looking to marry at this time, as I have other priorities.
I trust this will not cause any awkwardness between us.
Your friend in the Lord,
Miss Mary Anna Bumby

I hope this is not too abrupt or discourteous, while at the same time making it clear I am not interested.

I have no one I can discuss this with. John and Mr Hobbs have taken my only companion, Jane Hobbs, far from me. If John were here, I could at least talk about it, but it is unlikely I will receive any word from him before he returns, nor do I know where to direct a message.

One night I wake from a nightmare and think I see John standing by my door, but when I look harder, he is gone. I can still hear his voice, calling my name. It unsettles me greatly.

It seems marriage is in the air, despite the cold and rain. A canoe overturns just in front of the station with twenty people in it, on their way to a wedding. All the feast is lost in the water, although I am grateful to learn that everyone aboard makes it to land safely. The next week one of the influential chiefs of our area comes to the mission to get married. He brings with him a thousand supporters to partake of the feast. I'm glad I don't have to organise it, although I make him a cake with my honey as a gift.

One dismal afternoon, as I'm sitting mending by the fire, my thoughts drift back to the only other proposal of marriage that I've ever received.

Father's cousin John Bumby comes calling. Although he shares Father's name, it is not difficult to tell them apart. He is fifteen years younger, being the youngest son of yet another John Bumby, Father's uncle. He is also less fit; in fact, inclined to corpulence.

We share our condolences. His are offered for the death of my sister Jane Elizabeth. She was only sixteen and he did not know her well. Father doesn't always have much to do with his cousins, apart from the usual business dealings.

In return, we offer our apologies for the death of his wife Mary, who used to be Mary Pierson. She was 37, so her death was also unexpected. He is left with their three young children to raise.

Mother is never very friendly with Cousin John, because he is the innkeeper at the White Swan Inn. She does not allow any sort of alcohol in our house, and frowns upon those who use it. However, for politeness' sake, she entertains him and enquires about the children. One of his sisters is looking after them at present.

Then he says the most shocking thing. 'I wonder if I might start courting your daughter, Mary. It is time I found another wife.'

Mother doesn't know what to say. He is family, after all, and in need of assistance. It would be the charitable thing to do. However, he is much older than me, nearly forty to my twenty. And I am not sure that I want to become stepmother to his children, or indeed marry at this age.

He is not of our faith, and might be difficult to bring to our beliefs, given his age. If only he had followed his father into blacksmithing, like his other brothers. However, the drinking is the biggest stumbling block to his proposal.

Mother solves the problem for me, though. 'I am sorry, Cousin John,' she says, 'but I need Mary with me at present. I am not well, low in spirits with a troubling affliction, and I need her to stay at home with me.'

I feel such joy at her words that I'm sure it is the right decision. Not that I'm happy about Mother being poorly, which she is – she would never lie. No, it's her telling Cousin John that I am necessary to my family. Without Jane here, my duty is even clearer. I must assist my family wherever possible.

One Monday morning in mid-July, a letter arrives from Reverend Taylor of the Church Missionary Society in Waimate. I met him at the time of Governor Hobson's visit. I open it eagerly, hoping it holds news that John is on his way.

I scan the letter, hoping to glean the long-awaited information. However, the words that leap out at me are *condolences*, *regretful* and *sad loss of your brother*.

It seems that the writer has heard news of my brother, while it has yet to reach me directly. That John has drowned. Can this be true, or is it merely a rumour? How can I not have known? Although, I have been fighting a dull feeling ever since he left.

I'm plunged into utter distress, unable to eat or sleep. I agonise to know the details of what caused John's death. Will they be able to retrieve his body? It is as I feared, and my mind draws ever more graphic details. Has this long spell of inclement weather finally wrecked his fragile constitution? Or has his missionary zeal provoked him into a dangerous situation? I think back to the death of Reverend Williams, and my concern that cannibalism might still exist in New Zealand. The thought is almost too much to bear.

A messenger is sent to Mr Hobbs, who arrives the next day with Mrs White.

'We will send someone at once, to discover more details,' Mr Hobbs says, trying to console me. Mrs White is wrapping blankets around me, as I'm shaking uncontrollably.

'How is it that Mr Taylor knows, while we have not had any word?' I ask. It seems unfair to know so little.

'Shush, dear. I know it is hard to lose someone so beloved. Lord, I lost enough children to understand that. But we must look to His grace and mercy for our comfort,' Mrs White says.

Hemaima comes to me, tears streaming down her cheeks. She holds my hand and speaks to me in the native language. I fight through my anguish to understand her words.

'Kua mate te pumipi,' Hemaima says. The Bumby is dead. 'Ka noho ahau ki to taha ka hoe.' I will sit beside you to paddle.

'E mihi ana ahau ki a koe e Hemaima,' I say, with my eyes closed. The girls will watch over me and stay with me in my grief. If there is one thing the natives are good at, it is expressing their feelings. I wish I could lament the way they do, letting all my distress fill the air.

Later, Mata comes in. 'Let sorrow cease, and you be strong in God,' she says in her language. 'We will see him in Heaven.'

They are good girls and have taken our Christian messages to heart in a way that shames me. I should have their faith.

We wait the whole of the following day, but still there is no word. Finally, a messenger arrives with a letter from Reverend Fairburn in Maraetai. He says that John visited him and asked his advice on how to return to the Hokianga. John had been away from Mangungu for five weeks and was anxious to return. Mr Fairburn suggested an overland route, but John was footsore and weary, so decided that he would go by sea. About twenty natives accompanied him.

Mr Fairburn now regrets that he didn't lend John his boat. At the time, he hadn't trusted the natives to look after it. Instead, they went by canoe to Waiheke Island, where one of the natives had family. A larger canoe with a raupo sail was lent to them, and they left to sail up the coast to the Bay of Islands. Mr Fairburn saw them leave and felt full of foreboding.

It was several days before the six survivors of the sinking were able to get word out. John's body had not yet been found, Reverend Fairburn said.

I can't bear to think of him lying alone at the bottom of the ocean.

'I will go over to the Bay of Islands and hope to find a ship to take me down to the area,' Mr Hobbs says. 'If our dear Mr Bumby's remains are found, I would like to bring them back here.'

'I will go with you,' Mr Smales offers. 'He was such a great man; it is the least I can do.'

'I would appreciate that,' I manage to say. Then I think of something more. 'If you are travelling to the Bay, I must rouse myself from this anguish and write my letters home. Oh, how will I tell my dear father of this tragedy? And John's special friend, Miss Sells? And yet it is my duty, perhaps one of the last I can do for my brother.'

A service is held for the safe travels of Mr Hobbs and Mr Smales, for the soul of John, and for the ease of distress for his family and friends. It does little to comfort me. Feeling physically ill, I retire to bed.

John Hobbs and Gideon Smales set out on foot, taking some of the native men with them. I wish I was well enough to travel with the group. I want to see the things John saw, go where he had been, meet the same people. I feel more useless than I have ever been. Even supervising the house, deciding on meals, and entertaining visitors is beyond me. Without John to look after, I have no purpose. I wish I'd sunk on the boat with him, although I know that type of thinking is a sin.

The girls and Mrs White try to comfort me. They bring small delicacies to tempt me – mussels from the shore, pickled eggs, fresh cream. Though they are kind, I cannot bear their company for long. Mrs White prattles on while

I sit dabbing at my swollen eyes, wishing I could retire back to bed.

Physical pain attacks my body. My limbs are heavy and bumbling, my stomach and chest ache from sobbing, my face is puffed from tears. I feel that only God can understand my feelings, and I am grateful that I have that consolation at least. John will never come home to me, but one day I will go to him.

Only one of the group who left with John has survived. He makes his way back to Mangungu so that he can tell me what happened.

'The boat was big, big, long, filled food and blankets and books. We go Waiheke motu to Motutapu to Rangi-toto, and go for Kawau. Some want raise sail for go fast, but many stand on same side and boat go over. Many good swim, and we hold Mr Bumby for boat is up. We get him on and take wet clothes off. I hold him bottom of boat and others take out water.'

'But then you should have been saved!' I exclaim. 'What went wrong?'

'The sea big, big waves all over. Too many waters in boat, and boat over. And son of chief gone, we look for son. I try hold Mr Bumby, but he not help, and he go down.'

'Did he say anything? Was he scared?' I know that John was afraid of drowning, as he couldn't swim.

'In bottom of boat, he say, 'Ka Mate, Ka Mate' and more. He pray,' the man says. 'I save some book.'

'Thank you. They will be all the more precious because of the care you have taken,' I say. 'I will cherish all those things that are left to us.' The thought of his books and writing makes me sob even more. The last items he touched, a quill in hand upon paper.

I don't know whether it is any easier to accept John's death now that I have the details. I picture him in the bottom of the canoe, shivering with cold, saying his last prayers. I hope he was peaceful, knowing that the end had come. He'd often say that he would welcome death, in that it would bring him to the Lord.

I am in my brother's room looking at his things, holding his books and smelling his old shirt. It has taken a while for Mr Woon to find a way to unlock the door, as John took the key with him.

If only John could walk into the room right now. I miss his kind smile, his distracted way of listening to me. It seems incredible that I will never see him again in this life. He had so much that he wanted to achieve, and so much to offer to others. His faith was strong and carried me along. It is hard to believe that this is the Lord's work, but for John's memory I must keep trying.

Mrs White returns to her family. The worst of the shock is over, although some days it doesn't seem so. It feels as if my life has been divided in two; before I learned of John's death, and now when I am completely alone. It brings back the agony of my sister's death, and Mother's. Yet then, I had the support of John and Father. Now Father is thousands of miles away, still ignorant of the news, and John is in his watery grave.

The man who had tried to save John brings the boxes of papers he saved from the capsize. I set them around the fire to dry out, weeping all the while. They are in surprisingly good order. I am grateful that he took such good care of them. Some of the papers might be useful for a tribute to John. I'll find out which of his friends might want to write it, to preserve the details of his work.

A letter arrives from Mr Smales.

Kororareka, Bay of Islands
15 July 1840

My dear Miss Bumby,

I regret to inform you that we remain at the Bay of Islands, as we have yet to find a ship going in the right direction. The weather has been stormy, and all ships remain in port.

Please be assured that Mr Hobbs and I will hasten down to the Waitemata Harbour at our earliest opportunity. I imagine our chances of recovering your brother's body lessen every day, much to our consternation. However, we will do our best.

Mr Hobbs and I met Governor Hobson and his wife on the beach here at Kororareka. The governor sends his deep sorrow at the news of your brother's passing. He is also planning to travel south, to the same harbour, thinking it a better place for a capital.

While my task is one of sadness, I appreciate the chance to see more of this beautiful country. I only wish it was not an occasion of such unhappiness for you.

Yours in Christ,
Gideon Smales

Finally, more than two months after leaving, Mr Hobbs and Mr Smales arrive back from Thames and Waiheke without any good news. Mr Hobbs leaves straight away for Pakanae to see his family, but Mr Smales comes to visit me.

'Everywhere I went, people were upset to hear the news of Mr Bumby's death,' he tells me. 'He was highly respected in this country for his work. Even the CMS missionaries offered their sincerest sympathies.'

'As should be expected, for we do the same work of saving souls,' I say. 'Where did you go?'

'Mr Hobbs and I travelled to Waiheke, from where Mr Bumby had left. The people were full of sorrow at our loss and their own, for some of the missing natives were from there. I left them some gifts to show our gratitude for their care.'

'I should have thought to send some with you!' I say. 'I was told they had the son of a chief on board. Was he saved?'

'No, unfortunately. And Mr Hobbs was most distressed to learn that his Tongan man, Joel, was also lost.'

'I thought he was returning to the Islands with *Triton*.'

'He'd changed his mind and decided to come back to the mission with Mr Bumby.'

'Oh, that is very sad.'

'We also visited the Reverend Fairburn, who was the last Englishman to see Mr Bumby. He says he'd tried to induce your brother to travel overland, but that the journey seemed too strenuous for him. He was full of regret that he hadn't lent your brother his boat.'

I try not to let my resentment show on my face. It is all very well for the Reverend to have these regrets, but they don't bring my brother back to me.

'Thank you, Mr Smales, for your kindness in going on the search, and for being my envoy.'

'You are most welcome, Miss Bumby. Any time I can be of service, you have merely to ask,' he says, with a small bow.

Miss Bumby's Mission

CHAPTER 15

Spring has arrived again, the second spring that I've seen whilst in this country. Despite my recent neglect of my hives, the bees are doing well, building in numbers and bringing in pollen to feed the newly hatched larvae. I order several new skeps from Auntie, anticipating that I might need to split the hives to prevent the swarming that was so common last year.

The best-performing hive last season was the Sophia. In March, as the weather cooled, I'd again taken the duplet from above the main hive and drummed the bees from it into a basket. Then I'd shaken those bees back out in front of the main skep. Most of them had walked back into the original skep. Then I'd harvested the remaining honey and wax.

Now, as I prepare for the new bee season, I study my remaining jars of honey, carefully set aside for John's breakfasts. It seems unfair that the honey is waiting for an appreciation which will never come again. John often mentioned how grateful he was for the effort that I went to in preparing his sweet treat. I repeat his thanks to the bees every time I visit them.

It almost feels as if I should abandon my efforts, but I feel an obligation to the bees. They, like me, suffered the long journey on the seas to be here. They are God's creatures as well and deserve a chance to live in this wild country. And from the reaction of Heni and Mata and Hemaima, the natives will be happy to trade for honey.

It calms me to stand watching the bees as they return to the skep. If the day is fine, I can easily spend an hour watching as each bee flies unerringly to the skep, lands awkwardly and then totters towards the entrance. The colours of the pollen sacs on their hind legs are vivid and varied, from pale cream to bright orange to dull red, showing the range of plants that they are collecting from.

The bees do not lie in bed suffering when one of their number is accidentally swatted by a nervous native or fails to return from a collecting trip. They carry on for the sake of the rest. I take heart from their endeavours and resume my household duties and my teaching. There are still occasions when I break into heart-rending sobs, but the girls learn to settle me into a chair and bring cups of tea until I calm again.

The mail service with the Bay of Islands that Governor Hobson instigated brings news that a new capital has been established. It is on the shore of the Waitemata Harbour, not far from where John drowned, and is to be named after Lord Auckland, Hobson's patron. The Governor himself is still in the Bay of Islands, recovering from a stroke.

The mail service also brings letters from Miss Sells. It pains me to think that they were sent while John was still alive. The distance from England is so great that I have already been mourning him for months, but the news will not yet even have reached home.

Then, a month later, the letters I wrote to my father, Emma Sells, and Sophia Hyde are returned to New Zealand from Sydney. They have not found a ship to travel on, even after three months. I'm very distressed, knowing that our loved ones are still ignorant of all that has occurred. I hope that the news of John's death will not reach them circuitously, as it did for me with the letter from Reverend Taylor. I know how that uncertainty crippled me. The full knowledge of what happened is upsetting, but the imaginings I entertained were worse.

As the weather improves, so does my health, but only gradually. I am still very ill whenever I venture out on a boat, which confines me to Mangungu. Travel on the river is preferable to walking, as the country is generally untamed. Many creeks flow into the river, and dense stands of trees make the paths winding.

More letters arrive for me as the news spreads among the missionaries. It comforts me somewhat to realise how John impacted on the lives of others. In particular, Mr Whiteley from Kawhia says he longs to come to Mangungu to see me, but he is waiting for *Triton* to return with Mr Waterhouse. Then a few days later Mr Waterhouse writes to say that he's heard the news and finds it overwhelming. He has been like a second father to John, and integral to John's decision to come to New Zealand.

The constant reminders, though, are also like a poison, bitter and sorrowful. I try to carry on, going down to Pakanae by boat for a short visit with Jane. When it all gets too much, I retreat to John's room to weep, unseen by any observer. At times I feel as if he is there, and his room becomes a sacred memorial.

In mid-December *Triton* arrives back at the Hokianga, and I am delighted to see that both Mr Waterhouse and Mr Whiteley are on board. There are many tears at our reunion, such a sad contrast from when they left in May.

Mr Waterhouse has meetings with many of the district's missionaries, but he asks for some of my time.

'Miss Bumby, you know how I loved your brother, and how much I admire your faithfulness to him and our cause,' he starts.

'Indeed, I know that Brother returned your regard,' I say, tears springing again to my eyes.

'Unfortunately, my dear, the time has come when we need to determine your own future,' he says. 'The Mission House has been your home now for nearly two years, but we need it to accommodate missionaries. We have appointed Mr Hobbs as provisional chairman of the Northern District and Mr Whiteley for the Southern District.' He pauses, as if trying to find the right words. 'Tell me, what are your plans?'

I am shocked. This was home to my brother and all his earthly possessions are stored within its walls. Where will I go if I can't be here?

'Do you mean I should return to England?' I ask. 'I long to do so, for I miss my friends. Yet I feel that I have made good progress in learning the native language and have proved my worth with my school for girls. Also, I do not feel strong enough to endure another long sea voyage.'

Mr Waterhouse nods. He saw how the voyage out affected me. He also knows how devoted I was to John. He strokes his sideburn.

'I know, I understand, and I do appreciate all that you have done. Another possibility has been mentioned to me. I hear that young Mr Smales has offered his hand to you?'

I blush. Out of respect to Mr Smales, I have not mentioned his offer to anyone. I was waiting to discuss it with John on his return, even though I wasn't inclined to accept the offer. Then the disaster swept it from my mind for a long time. It can only have been Mr Smales who discussed it with Mr Waterhouse.

'Yes, but that was six months ago,' I say. 'He may have changed his mind.'

'I believe not,' Mr Waterhouse says, smiling. 'It does sound like a reasonable arrangement, doesn't it? He is an eager and energetic young man, and it seems that he would welcome your companionship.'

I think back to his readiness to search for John's body. I also remember his easy company, and the thoughtfulness and faith that show through his sermons.

'I suppose it would be a good thing to do,' I say. 'I've been very lonely here without my brother. And that would allow me to continue my work.'

'Excellent. That's decided, then.' He claps his hands together.

'No, indeed, I must pray on it first, please. As you can see, I'm still in mourning. Would it be right to marry so soon? And it is a large undertaking, as I never expected to become a wife. Can I convey my decision tomorrow? And in fairness, I think I should inform Mr Smales first!'

I spend another sleepless night in contemplation. I don't feel that I have much choice. The Society doesn't value my work as being meaningful like the men's, and doesn't expect to support me in it. I can return home, perhaps, but I still feel weak from my mourning. I'm not sure I could survive the voyage.

The next morning, I seek out Mr Smales, and find him in the printing room. Mr Woon is tapping letters into blocks while Mr Smales is sewing portfolios. I suppress a smile at seeing a man with a needle in hand. His fingers look too large to hold such a delicate item.

'Mr Smales, I wonder if I might have a word with you?' I ask.

The mission is busy with all the visitors and preparations for the Christmas service, only a few days away. To gain privacy, we take a walk into the forest where it is cool and dim.

'How can I help, Miss Bumby?' Mr Smales asks, as I am struggling to know how to begin.

'Many months ago, you asked whether I would become your wife. And I was wondering whether the offer still stands.'

I chance a sideways look at him. He really is quite handsome, vigorous with dark hair and a strong chin.

'Most certainly. I have not changed my opinion of you, Miss Bumby.'

I feel weak, a tremor passing through me. I can't believe I've been so forward, or that this is really happening.

'That is most gracious. At the time, I felt I did not know you well enough. But now I have observed you and seen your commitment. Together, I believe that we can achieve wondrous things.'

'Miss Bumby, you bring me great joy. I have seen your suffering at the loss of your brother, and have longed to console you, but it has not been my place. Thank you! I will strive to be the best husband you could want.'

'And I pray that the Lord will help me discharge my duties as a missionary's wife.'

Mr Smales seizes my hands and plants kisses on them. I resist the urge to pull them away. 'How soon can we be married?' he asks.

I feel it's all proceeding too quickly. 'Please, Mr Smales. I am still in my mourning clothes for my brother. Surely there is no urgency? What will people think?'

'I apologise, Miss Bumby. I don't wish to diminish the depth of your loss. Yet, he was a brother, not a husband. Do you not think he would be happy to see you established in a household?'

'That is true. And I admit I will welcome the chance to cast off some of these black clothes, now that summer is with us again.'

'Do you think we could ask Mr Waterhouse to officiate?' Mr Smales asks. 'I know he is like a father to you.'

'Yes, and he is due to leave again shortly,' I remark. Mr Waterhouse is a link to the past, and someone I greatly admire. 'Perhaps then we should marry in one week.' I feel I am caught in a river current, being pulled along without hope of rescue.

'Mary. May I call you Mary? And you must call me Gideon. Mary, you have made me very happy. Before this year is out, we will be wed. I have dreamt of this. And prayed, of course. Bless you.'

Gideon's enthusiasm is infectious. I try to shrug off my remaining reluctance and give him a smile. Perhaps the life of a missionary wife will not be so bad if I have such a loving companion beside me.

I allow him to hold my hand as we walk back to the mission to find Mr Waterhouse and announce our plans. It feels bold, even audacious, to be touching so intimately in public, making a statement to the world. Perhaps his is the hand of rescue.

Mr Waterhouse, Mr Whiteley, and the Hobbs family are all at Mangungu for Christmas and the wedding, as well as the Warrens and Creeds and Woons. There are about twenty all gathered, which keeps me busy with meals and unable to dwell on my fate.

'Oh, my dear, I am so glad for you,' Jane tells me. 'It is a sad future for a woman without children.'

I can't meet her eye. 'Of course, we may not be blessed like that,' I say. 'I am getting rather old to have children.'

'Nonsense. I'm ten years older than you, and still bearing babies. And Mr Smales seems healthy as well as being young.'

'Yes, I hope I do not have to concern myself with his health, in the way I did with my brother,' I say, trying to steer the conversation onto safer ground.

'Now, my dear, did your mother ever talk to you about, well, marital duties?'

'I was never close to being married before,' I say.

'Well, I feel I ought to take the place of your mother, so that your wedding night is not a surprise. But you are not a city girl, Mary. You have seen our dogs mating, for example.'

Suddenly I understand what Jane is trying to explain. It isn't about having a hot meal ready at dinner, or ironing your husband's shirts, or agreeing with everything he says. This is about the most intimate matters. How babies are made.

'Does a man really mount his wife like that?' I ask.

'Usually, she lies on her back, and he puts his appendage into her from above,' Jane says, and covers her mouth.

'How embarrassing. I fear I won't be able to look.'

'Yes, the first time can be uncomfortable, maybe even a little painful. But sometimes you can get to enjoy it.'

'And you do this every time you want another child? Or just once?'

After all, the queen bee only mates with a drone on her marital flight, and then has eggs to last the rest of her life. But animals are mated when they are ready to have progeny. I wonder why Jane and Mr Hobbs don't stop doing it, as their family seems large enough already. Then I blanch at the image that thought provokes. I'll never be able to look at Mr Hobbs again.

'Oh Mary, haven't you been paying attention to the world? This is something men desire to do frequently. And some women. This is why whalers and merchants have native wives, and why ships call in to Kororareka. God says that it should only happen between husband and wife, but men will ignore that if they feel the need.'

Suddenly I think of Gideon. Is this why he is marrying me? So that he can do this thing to my body? I'm not sure whether I can bring myself to sacrifice my physical form to his desires. Maybe I can ask him to refrain from this act.

I consider all the other married couples I know. This is what Nancy must have done with her Davy. What the Woons and Creeds do in their beds. What any woman with a babe must have done.

Maybe it isn't so bad if everyone does it. It sounds revolting, too intimate and earthy, something no civilised person should do. Yet if it is consecrated by marriage, before the eyes of the Lord, it must be good. And the *Bible* does command people to be fruitful and multiply.

'Our little house in Pakanae is going to suit you and Mr Smales very well,' Jane continues, looking around the Mission House, 'while this house will be much better for our family.'

'What do you mean?' I ask. 'This is our home!'

'Well, yes, it has been. Perhaps I shouldn't have mentioned it, but I thought Mr Waterhouse had explained. You see, my husband has been appointed Northern District Chairman, and as such we'll live at the Mission House. Added to that is the fact that he built it himself. He's not very happy taking on the new role, as he prefers being outside and doing practical things instead of reports. But he knows it's his duty.'

'Oh. I wonder if Mr Smales knows this. So, we need to move down to Pakanae?' The settlement where Jane lives is closer to the heads, and further from the path over to the Bay of Islands. I'll also need to become acquainted with the natives who live near there.

'You have no need to worry about this now,' Jane says. 'My husband tells me we should give you a month after your wedding before we change residences.'

I've noticed that Mr Hobbs and Gideon are very stiff around each other. I wonder whether this move is intended to punish Gideon. Yet losing this house will affect me more. This house is where my memories of John are stored. His clothes, the cup he preferred, the desk where he wrote his letters. Leaving here is like abandoning his legacy. My life and those of everyone else are moving on and I fear that he will be forgotten. That his great sacrifice, his life, will be consigned to history as not being worthwhile.

I determine that I will never forget him. I feel sure that Gideon will help me in this. John Bumby will be remembered.

The day of the wedding is bright and hot. I adorn my best gown with a spray of the small purple flowers of the hebe. The little chapel is crowded with our well-wishers – the missionaries, my serving girls Hemaima and Mata, and the other native mission workers. Even Nancy and Davy have appeared, a bairn in Nancy's arms.

I stand at the altar with Gideon in front of Mr Waterhouse. I tremble as I repeat the vows, consenting in the name of the Father and of the Son and of the Holy Ghost to bear each other's burdens and to become one forever. I've never been at the centre of attention like this before, with all eyes on me. I wonder what they are thinking of me. Whether I've made the right decision. But now there is no going back; I have promised myself to another, in front of God.

It feels like a huge responsibility. Now, no matter what happens, our lives will be intertwined. I will learn things about this man, who is still virtually a stranger to me, which will surpass the intimacies I formed with my sister, my mother, and of course, my brother. My bed will not be my own, to cry alone in when things get too difficult. And how am I going to use the chamber pot at night with someone there to listen?

Will I get to share my thoughts and fears? And will he do the same? When John was alive, I struggled not to weigh him down with my concerns. Sometimes I wasn't successful, but I tried to contain my worries to myself. My diary was the place for confessions, not the dinner table. The troubles of the world and beyond weighed on John enough already.

What is Gideon really like? My husband. I glance at him, tall and handsome in his best suit. He seems confident and happy. He catches my eye and gives a wide grin.

The congregation begins the hymn, *Love divine, all loves excelling*, one of my favourites. I look at their faces. Two years ago, the only people here that I knew were Mr Waterhouse, the Creeds, and the Warrens, as we travelled on the *James* from Cape Town to Van Diemen's Land. Now I can see the broad face of Jane Hobbs, kind and familiar. The serious face of my husband, concentrating on the words. The giggling of Hemaima and Mata, almost overcome with the occasion of their mistress' nuptials. The fat faces of the Woons, and the thin balding one of Mr Whiteley. It is a different world I belong to now. That most intimate visage of my brother is missing. But the new profile beside me is one I will get to know even better. Gideon Smales, my husband.

I try my new title. Mrs Gideon Smales. No more jokes and sniggers about Bumby, although I feel disloyal to abandon the name. In Thirsk no-one thinks anything of the name Bumby. It is familiar, common, an old name rumoured to have been brought by the Normans. Now I'll rarely hear it. Since John's death, I've understood that my father's surname will not continue down the line. It has been another wound to bear, and I wonder how my father is coping with the news of John's passing. If he's even heard of it yet, though six months has gone by. He will be surprised to learn that I am a married woman. There has been no opportunity for Gideon to ask my father's permission, of course. Mr Waterhouse stood in his place, as he has ever since we left England.

And John. Is he looking down on me from Heaven? I hope he is pleased for me. I did my best to keep my promise to our mother about looking after him. If only I'd been able to persuade him not to go on that last journey. And Ma, and Jane – are they with John, all together, keeping me in their sight?

I put my gloved hand into Gideon's. It feels such a bold move, in front of everyone. He squeezes, and it reassures me. We'll go forward together. I'll devote myself to this man, as I have promised the Lord to do. I have purpose again.

CHAPTER 16

As the new year 1841 comes in, I look back on the previous year. So much has happened that I could not have foreseen. I still have a full household of visitors to cater for, as well as the strangeness of having a husband.

Gideon proves to be an attentive husband. He notices when I put flowers in a vase, when I sigh with exhaustion, and he enjoys being close to me. It surprises me, this feeling of unity. I've observed other couples; some, like Mr and Mrs Hyde, seem to respect each other and care deeply. Others, like Mr and Mrs Hobbs, sometimes appear to only tolerate each other. And still others obviously feel irritated with the shackles of marriage, where the husband is to be treated as the superior.

In our marriage bed, Gideon is gentle and patient and grateful. As Jane told me, he is interested in repeating the exercise. Yet he is also very understanding when I tell him I am sore. He apologises repeatedly and is content to hold me close.

There is much to learn. I am surprised to find that Gideon's parents were not married. His father, 'Owd' Gideon, had a long-term relationship with my Gideon's mother, Susannah Nares, and Gideon was the middle son, given his father's name. But after the partnership broke down, 'Owd' Gideon married and had other children, including a half-brother also called Gideon Smales. The Smales family have a very successful ship-building business in Whitby, and his father was willing and able to help my Gideon gain a good education. Despite his growing family with his wife, 'Owd' Gideon still took an active interest in his first family.

Meanwhile, Gideon's mother Susannah married a whitesmith, David Gaskin, who worked with tin, pewter and other light-coloured metals. Gideon was only four, and the brothers were quickly accepted by their stepfather. The Gaskin family had strong ties to the Wesleyan church.

Gideon remains very close to his mother. He writes to her of our marriage, asking for her blessing.

'You would enjoy meeting my mother,' he tells me. 'She is a strong person, like you. Able to cope with adversity, well-educated and adventurous.'

It amazes me that my husband considers me strong and resilient. I've spent so long mourning John and feel myself weak and feeble. I determine to live more closely to Gideon's expectations. Maybe I misjudged him when I was ascertaining his reasons for marrying me.

Slowly, our guests depart. The last to leave are Mr Waterhouse and Mr Whiteley aboard *Triton*, along with Mr and Mrs Creed.

'Are you well and happy?' asks Mr Waterhouse as he takes his leave. 'Guide your new husband well, my dear. I fear he may be overly confident and stubborn.'

He is referring to the disagreement we've experienced with Mr Hobbs. A letter from Governor Hobson to John went unanswered by me after John's death.

It embarrassed Mr Hobbs when he met the governor and had to admit to being unaware of the letter. He ordered me and my new husband that any mail addressed to John should be forwarded immediately to himself. Any personal correspondence would be returned to me. However, Gideon protested that I had the right to decide how the mail was dealt with.

It feels like spite when Mr Hobbs informs Gideon of the house swap – set for Monday 25th January.

All our household items are packed and ready for moving, but Monday dawns grey and develops into torrential rain.

'We should start getting the boys to move our things into the canoe,' I fret. It is a long journey, about seventeen miles towards the heads.

Gideon looks out the window at the sheet of water pouring from the roof. 'It's far too wet,' he says. 'We should wait it out and go tomorrow instead.'

'But Mr and Mrs Hobbs will be arriving here,' I say. 'It was agreed.'

'Ordered, more like,' Gideon growls. 'No, I'm sure they will have been sensible enough to delay their own departure.'

In the evening, though, Mr and Mrs Hobbs arrive on the doorstep, unsmiling.

'Why are you still here, Smales?' thunders Mr Hobbs. 'Can you not follow the simplest of instructions? This is now our house. Gather your things and get out!'

'You can't expect us to leave at this time of the evening!' Gideon replies. 'It wouldn't be safe for my wife to be out on the water in the dark.'

'You should have thought of that earlier.' Mr Hobbs is not in the mood to be conciliatory.

Fortunately, the native men we've employed to take us are still about. It takes many trips between the house and the riverbank to load everything into the canoe. Gideon settles me into position, then finds a place for himself.

I watch as the men take up their paddles and row into the current. I look back at the Mission House, my home for the past two years. I can see the lamplights through the windows as the new owners move their possessions into place.

When will I see it again? How can I hold on to the memory of my brother without its prompts? Even my friendship with Jane seems spoiled. Can I bear to visit, having to be hosted by the interloper?

As the paddles dip into the water and the canoe moves further away, I remember my hives. It is not a good time of the year to move them. They are full of honey and fragile with the heat of the bees, which softens the wax. I'll have to engage a native to bring them down to Pakanae in a few months. I wonder whether Heni will continue to care for them and whether Marianne will revive her interest.

As well as the house, garden and hives, I am leaving my girls. Mata and Hemaima belong to the local tribe and don't feel comfortable moving into a competing tribe's territory. Also, they've been with me for nearly two years. It's time for them to return to their families, find husbands and have children. I'm happy that they'll take their Christian ways and literacy skills with them, helping to convert even more people.

The morning sky is lightening behind us when the canoe finally reaches Pakanae. I am exhausted. The journey has been rough, with strong waves battering us and threatening to overturn the canoe. John's last voyage plays on my mind. Am I going to suffer the same fate? After all, I have my skirts to pull me beneath the surface, and no better swimming ability than John. Only three men have volunteered to paddle us, and it has taken nearly eight hours because of the load of goods in the boat.

When I am carried ashore, I fall to my knees and pray. I thank the Lord for our safe deliverance and ask Him to assist in our work of helping the poor natives to come to Him. Then Gideon helps me to my feet, and we walk up to our new home.

Pakanae is a poor substitute for Mangungu. The house is less grand, the land around it windswept and poor. There are only a few trees, insufficient to provide the protection afforded at Mangungu by its enveloping half-circle of hills and forest.

The servants that Jane Hobbs left behind are surly and unwelcoming. I suspect they feel abandoned. When I try to question them on their faith, they indicate that they are not willing to change their beliefs from their pagan ways. Their atua are more important to them than our English god, even though I try to emphasise that the Lord rules over all, not just the English.

I regret that Hemaima and Mata have chosen to stay behind.

The circuit of preaching which comes with Pakanae includes a village on the Whangape Harbour, about 24 miles north. The journey there is going to be arduous for Gideon. He'll have to cross the Hokianga River and then either take a long trek up the coastline through forest and sand dunes, or a sea voyage across the bar.

This station at Pakanae was established by John Whiteley and James Buller four years previously and named Newark by Whiteley. The only consolation I have is that Mr and Mrs Creed are also here.

'We get to meet many interesting people by being here,' Eliza Creed informs me. 'When ships arrive, they often stop here as their first landfall, after their voyage from Sydney or Hobart Town.'

'Is it easy then to send and receive mail?' I ask. I'm worried that I'll be cut off from communications, being much further from the overland route to the Bay of Islands.

'Oh yes, we are often the first to receive news from home,' Eliza says. 'And our people do a good trade, supplying pork and kumara. The ships have to travel further upriver to collect timber, especially kauri, but we have the flax fibre, what the natives call harakeke, which is proving useful for ropes.'

'Is it having a civilizing influence then, with the natives engaged in trade and work?' I ask.

'Unfortunately, they have taken well to learning the value of earning money,' Eliza replies. 'It can be difficult to find enough people to work for us, unless we match the wages the timber merchants pay them. And they are still keen to buy muskets and axes, no matter how hard we preach against violence and murder.'

'Then we must work even harder to bring them to our Lord,' I say. I determine that I will help Gideon all the more in his work.

Gideon is struggling to master the native language. He does not have the natural affinity with it that John showed and has failed to find a helpful acolyte among the tribe to help teach him. When he visits the villages in his circuit, he finds them unresponsive to his messages and sometimes even a little hostile. Instead, he stays home and concentrates on writing his sermons.

'Aren't you meant to be helping Mr Hobbs bind books?' I ask one morning.

'I know he expects it,' Gideon replies. 'But it is a long way to travel, and his reception will be cold. I prefer to concentrate my efforts around here.'

'Do be careful, husband,' I say. 'I think Mr Hobbs is not a man to cross.'

'I feel I am being careful. We do not deal well together when we meet, so I am avoiding that circumstance.'

Eliza is correct that the mail often arrives first in Pakanae. We are able to gather items addressed to John and then send anything official on to Mr Hobbs. In

March, a parcel of letters arrives. Gideon collects them, despite the fact that they appear official.

Mr Woon observes him take the parcel. 'Do you think you should send those on, Mr Smales?' he asks. 'They look to be the package we've been expecting from London, to judge by the writing.'

'I cannot tell,' Gideon says. 'They are addressed to my brother-in-law, and so are Bumby property. I will open them with my wife and ascertain the correct course of action.'

When he learns of it, Mr Hobbs travels down to Pakanae and storms up to Gideon. Speechless with rage, he presses a note into Gideon's hand. Gideon glances at it without reading it, and then throws it onto the table.

'I must insist that all mail addressed to John Bumby is given to me, as the Chairman of the District,' John Hobbs manages to say, through gritted teeth.

'On the contrary, as the closest relatives of Mr Bumby, my wife and I will continue to open any mail addressed to him and forward any official documents to you.'

Mr Hobbs takes a deep breath through his nose, and I can see his nostrils turn white. His eyes bulge.

'You will do no such thing. I am your superior and you will follow my instructions.'

Gideon clenches his fists and squares his shoulders. I fear that he will strike Mr Hobbs. Fortunately, he lifts his chin and issues a warning instead.

'I will follow no instructions that are clearly wrong,' Gideon fumes. 'And if you presume to open any mail addressed to Mr Bumby, I will insist that the magistrate prosecutes and fines you for interfering with the mail.' He picks up the note and stuffs it into his pocket.

'Oh, Gideon,' I say after Mr Hobbs has left. 'I wish you hadn't threatened him.'

'He deserved to be hit,' Gideon says, 'but I restrained myself. The pompous git. How dare he use his rank to force me.'

'Poor Jane. I wonder she can bear him,' I say. 'Still, I'm unsure whether this is an issue we should fight him about.'

'He's not showing any respect to you, as John's sister. That's what I can't abide,' Gideon says.

If Gideon thinks that his resistance has decided the matter, he is wrong. Mr Hobbs gathers other members of the Northern District Committee and presents his side of the argument. They agree to ask Gideon to attend a special meeting in Mangungu. Gideon ignores their invitation, despite my urging. A few days later they meet again, and Gideon arrives late.

'I apologise for my lateness, but I was hindered by weather and the natives,' he says, facing down the frowns of the men.

'I put it to you, gentlemen, that we should decide whether Brother Smales' apology is satisfactory or not,' Mr Hobbs says. The glint in his eye does not bode well for Gideon.

'I agree. All those in favour of accepting the apology?' Mr Woon says. No one answers. 'Against?' All four men raise their hands.

Gideon pulls his shoulders back. He expected as much from Hobbs, but he is disappointed that the others have sided against him.

'Very well then. I have here a letter from my wife regarding the documents and mail. May I read it to the meeting?' My letter is meant to mollify, and explain our position.

'No, I don't think that is necessary,' Mr Woon says, the others nodding their agreement.

Gideon bites his lip in frustration. It seems that the committee is united in their position. What can he say to bring them alongside?

He tries to speak about the work he's been doing at Pakanae but is met with stony silence. Finally, he asks, 'Where do you consider I have erred?'

'Your threatening language to me, stating that you would go to the Magistrate and prevent me receiving any more letters of the kind, induces me to say that by so doing, you endanger the interests of the District and your own character, in the estimation of the committee,' Hobbs says. He has a gloating smile at the corners of his lips, as if he believes he has won.

Gideon leaves the meeting unrepentant, despite the committee's stern reprimands. The brotherhood of the District appears to have deserted him, even though it seems to be the result of a vindictive strike for personal reasons from Hobbs.

I listen to Gideon's recounting of the meeting with dismay. I've been feeling poorly again, despite not having been back on the water since our arrival. It seems that my ill-health is a penitence I must bear.

'And then, they had the temerity to bring up the matter of your brother's estate!' Gideon exclaims through his mouthful of dinner.

'What has that to do with this business?' I ask.

'There were some monies owed by Mr Bumby to the District. They say that, as we have inherited his possessions, we must discharge his debts.'

'Do you have a copy of his account?' I ask. I take it and study the figures. 'No, this is not quite right,' I say. 'They have not allowed for his clothing allowance, nor for his payment for the ship which took them south. We must write to them and get it adjusted.'

'Yet another way for Hobbs to persecute me,' Gideon says. 'The spirit of the Lord seems far from him.'

Gossip and rumours start to spread around the district. Gideon is not shy about telling his side of the story, but his version doesn't reach the ears of the senior officials in the Society. They send sternly worded letters of rebuke. I write to Mr Waterhouse, hoping to persuade him to support us.

The native Christian converts are confused by this rift between the Wesleyan missionaries. It seems at odds with the sermons we preach. They are also struggling with dissension in various tribes, as the tohunga or native priests have started to resist the influence of the Christian message.

The local natives are not as friendly as they were previously. I make sure to lock the door at night and try not to think about the massacres of earlier times. Then word arrives from further south, that one of the chiefs has ordered that Gideon be shot. We don't know whether it is related to the discord within the Society, or part of a general unease about the continued incursion of outsiders.

It is a worrying time, as I fear that my husband might not return from his trips. The anniversary of John's departure is close, and I think about my sense of dread when he left, never to return. I find excuses to keep Gideon at home.

'Do you think they will shoot you?' I ask, one night in bed. 'What can we do to protect you?'

'I can't judge how serious the threat is,' he says, shaking his head. 'I do wish we were closer to Mangungu, though. Being near Nene's kāinga always felt safer.'

'Will his protection extend down here?'

'We can only pray that it does. For both our sakes, and that of our babe.'

For by now it is evident that I am with child. I am grateful that the illness that plagued me is now receding. Gideon is delighted at the news.

'My father's first grandchild,' I say. 'And I am following in the footsteps of the Queen. If it is a girl, perhaps we should name her Victoria, like the Princess.'

'It will be a boy,' he says. 'I'm from a family of boys, all strong and healthy. This is truly a blessing from the Lord.' He puts his hand over my belly.

'You must stay alive for him, then. A son needs a father.'

I send a letter upriver to Jane, asking her to arrange for the hives to be brought to Pakanae. In reply, I get a letter from Marianne, the handwriting careful and rounded.

Mangungu Mission House
26th April 1841

Dear ~~Miss~~ Mrs Smales,
How are you? Mama and Emma and Retta and Richard and Sophia all send their regards.
Heni and me have been carring for the bees since you left. They are all winterd down good and come out on sunny days. We think it would be good if you left some here at Mangungu. Pakanae is rather cold. But when I lived there, I missed them and I think you miss them. Can I send two new hives?
I wish you still lived here. I have lots of questins. Papa says he will help me also, as he is intrested in bees from seeing me do it.

Yours in Christ,
Marianne Hobbs
P.S. Edward is now walking.

I am charmed by the letter and send off my reply immediately.

Dear Marianne,
I am relieved to hear that you have been looking after the bees, with Heni. I think you will both be wonderful beekeepers. You may have plenty of questions, but if you observe the hives carefully and think of what is best for the colony, then you should make the right decisions.
Your suggestion of sending two hives is very sensible. I started with two when I first arrived, and it will be less unsettling for the rest.
I will send one of my men, Wiremu, to help Heni bring them here. I hope Wiremu will be as brave around the bees as you are.
I miss you and Mangungu and your family also, but we must accept our roles, as the bees accept theirs.
Maybe we can continue to correspond, so that if you are unsure of anything, you need only ask.
Your loving friend,
Mrs Mary Smales
P.S. I nearly wrote 'Bumby' there. It is a difficult thing to change your name!

I am surprised to hear that Mr Hobbs is paying attention to the bees. He was never very attentive when I was first teaching Marianne. And I can never quite forgive him for his remarks on my surname, or his treatment of my husband, even though that would be the Christian thing to do.

I am just beginning to need to adjust my clothing for my increased girth when I am surprised by a visit from Lady Franklin.

'My dear Miss Bumby – or no, I must call you Mrs Smales now. I wept when I heard the news of your poor brother. But how fortunate you have been to find a husband.' Lady Franklin's gaze sweeps over my poor excuse for a parlour. I hope my new girls will serve the tea as I've shown them.

'Thank you, Lady Franklin. How is Sir John? He is not with you?' There is a small party with Lady Franklin, but no sign of the jovial, portly Governor.

'No, unfortunately my husband has had to remain in Van Diemen's Land to conduct affairs of state. But I've been able to travel around this magnificent country you live in, and meet some wonderful people.'

'I have not yet had the opportunity to venture far from here,' I say. I haven't much ambition to do so, but Lady Franklin is such a reputed adventurer that I hesitate to admit it. 'Tell me, where have you been?'

'We travelled down the east coast, calling in to Tauranga, and then further south. The middle island is particularly beautiful with many snow-capped mountains. It reminds me of Van Diemen's Land, although the green is of a different hue. And on the way here we stopped off at the south heads of the Manukau Harbour, to visit some settlers. They have been living such a harsh and frugal life. I intend to send one of my dresses to a settler woman there. They say they will name the area Franklin! Such an honour for my dear husband.'

I feel that I am also leading a frugal life, so perhaps Lady Franklin might like to send me a dress. The Society has sent word that they are heavily in debt, and that the New Zealand missions' future is in doubt. No further drawings are allowed for the next six months. I am fortunate that John's estate provides us with some discretionary funds, so that we are still able to serve cake and proper tea, not the weakened version of a second brew.

My girls are kept busy preparing meals for the visitors, although Lady Franklin's servants do provide some help. I'm inspired by this great woman, whose devotion to her husband is so evident. I wonder how she can bear to be apart from him for so long. Already I crave the company of my own husband and feel any absence acutely.

What a shame that Lady Franklin has never had children. She does not mention my condition, and I'm not sure whether it is from disinterest, propriety, or a deep-seated sorrow, so I don't mention it either. It reminds me of how I'd felt before marriage, when the idea of being with child was scary and somehow revolting. Now, it is exciting, the idea of Gideon and me creating a new life. I can't wait to see Gideon as a father.

After dinner, Lady Franklin and her companions set off on a walk to explore the area. She hurt her leg quite badly when she visited the French settlement of Akaroa, so is carried in a litter by a group of native men. Even so grievous an injury doesn't keep her confined to home. I beg off, citing tiredness. It is true – having to be on my best behaviour for such a grand lady is tiring, and I must supervise the girls to provide supper.

The following day their party sets off back to Mangungu and we return to our quiet life.

CHAPTER 17

Wiremu and Heni arrive with the skeps in a canoe. Heni has remembered to block the entrances with grass, and only a few bees escape and fly around the hives as they are unloaded and taken up to the mission.

Pakanae should be far enough away from their home site for the bees to settle down. I remember when my old bee mentor, Mr Webster, sold a hive to a man from Northallerton, eight miles away. Unfortunately, he opened the hive when he was only partway home, and many of his bees returned to Thirsk.

I have tasked Wiremu with making a shelter for the hives. I've shown him how to make a dry-stone wall, sheltered from the prevailing westerly winds, and how to leave some spaces within the wall for the hives to sit in. Boles, they are known as at home, rather than holes. They keep the skep drier, lifted off the ground. I've had two winters in New Zealand now and know how wet the lashing rain can be.

Wiremu is obviously very nervous about handling the hives. I've made him a veil and found some gloves, but he still hangs back as Heni and I take the skeps from the half-barrels and place them in the boles.

One of the hives is roaring.

'Can you hear that, Heni? Whakarongo mai. The bees are angry, rikarika.' It is markedly different to the other hive, which has a faint buzz.

Heni nodded. 'Ae. He aha ai?'

I shake my head. I'm not sure why they are upset. Perhaps some of the comb has broken during the journey. I've only heard bees roaring like this once before. Then, it was because their queen died, and they hadn't yet made another.

'We'll leave them here for the night,' I say. 'They should have unblocked their entrances by then and might have settled down. Ka haere tatou.'

Wiremu is only too happy to retreat further. I watch him, jumpy at every puff of wind, thinking that the bees are coming for him. He is no replacement for Heni, who now works the bees without gloves. I'll have to find another worker to help, as I get bigger with child.

As it turns out, it is one of my kitchen girls, Hera, who takes on the role of assistant beekeeper. She is a surly, indifferent kitchen maid, but is strong and enjoys the outdoors. Hera is fascinated by the activity of the bees. She's also excited to be given a veil and gloves, and to be encouraged to smoke, so that she can blow into the hive entrance and calm the bees down.

One of the hives is still upset when Hera and I go to check on them.

'No kuini,' I tell Hera. 'Ka mate te kuini.' At least, that is what I suspect – that the queen is missing, and the workers are unsure what to do.

'Kua mate a wikitoria?' Hera asks, confused.

'No, not Queen Victoria,' I laugh, although Wikitoria might make a fun name for this hive, if it survives. 'Te pi kuini.' I point to the hive. I'll have to start again with explanations and teaching. At least I have better language skills this time, since I don't have Marianne on hand to interpret.

I get Hera to hold the other hive high, so that I can reach underneath and break off a piece of the comb. While Hera lowers the skep back onto its base, I look into the hexagonal wax chambers. As I hoped, there is a tiny egg at the bottom of some of the cells, and white splodges in others. Eggs and young babies, the brood.

'Whanau,' I tell Hera, showing her. I pat my own stomach. 'Tamariki.'

Hera looks at me incredulously, then looks again at the honeycomb. 'Tamariki?'

I nod. I agree that it does seem strange. The babies look nothing like bees. Even at an advanced stage, they are more like those huhu grubs the natives eat. White and amorphous and passive, sitting in their cells waiting to be fed. Then, when they fill the entire cell, the worker bees put a wax capping over their heads and the grubs change. When they are ready, they chew a hole in the lid of their chamber and clamber out. I will have to show Hera that process. It always fascinates me to watch an emerging bee. It is like watching a chicken hatch. One minute they are only potential, and the next they are a fully formed individual.

Is that what it will be like when my bairn arrives?

I point to the other hive. I hope my veil and gloves are secure, without holes. These bees are going to be annoyed.

Hera picks it up, and bees stream forth. I am amazed and grateful to see that Hera doesn't even flinch. I quickly put the comb up inside the skep, pushing a few thin sticks into it to hold it in place. Then I motion Hera to put the skep back down. We move away from the bees that are bombarding us.

'Kuini hou,' I tell Hera. I hope that the hive will make a new queen from the eggs and larvae I've just inserted. If the hive has been queenless for a while, all their eggs and larvae will have already grown. The bees will feed these new eggs differently, with copious amounts of royal jelly to make the larvae bigger and fertile. The bees build larger chambers for the queen babies to rest in until they are grown. Then all I need to worry about is that there are drones around to mate with the queen.

I have to try something. Otherwise, the whole colony will die out. That is the responsibility of the beekeeper – to provide the right conditions to keep the hive alive. Each individual isn't particularly important, but altogether they make a cohesive society.

Plenty for us to learn from, I think. Maybe by teaching beekeeping, I can show the natives how to live more harmoniously with one another.

News of John's death has now reached England, and packages begin arriving from my family and friends. I am delighted to see *James* return to Hokianga, although with the wind so favourable, the ship sails past Pakanae to Mangungu. Gideon organises a canoe and paddlers, then sets off after it.

He returns the next evening with news that Captain and Mrs Todd are on board and hope to spend some time with me. He also brings a box from my father, but has had to leave other boxes from Mrs Hyde and Miss Sells on board until they are uncovered from the cargo.

The letter from my father shows the mark of his wife, Elizabeth, in its phrasing, but the sentiment is all his.

My darling daughter,
What devastating news from you. I could scarce believe it. Our John, so certain of his calling, perishing in the wilderness with only savages for company. I have been in shock these last few days, wishing I had forced him into trade, rather than allowing him to follow his passion. Yet it

so pleased his mother that he was serving the Lord. I could not have done otherwise than send him to Leeds.

Now I put pen to paper to urge you to return to us. We are your family, and there is a place for you here in our hearts and at our hearth. Do not remain in that foreign land that has stripped us of so much that is worthy.

As you say, our consolation is that he will be in Heaven with his mother and sisters. Of that there can be no doubt. He will be well deserving of his reward.

I hope that our next correspondence will be in person, as I welcome you home once more.

Your loving and faithful father,
John Bumby

All my mail is addressed to Miss Mary Bumby, of course. The word of my marriage will only just be arriving in England now. How surprised they'll all be. I didn't drop a single hint beforehand, obviously. They might be looking for my arrival on every ship. Once, it seemed my only option. How glad I am that I took the opportunity to marry Gideon. He has become my saviour, in ways which restore my belief in the mercy of God.

From Hokianga, *James* is heading to Sydney. I have a surge of energy at the thought of contact with my friends there and write letters to them all. Then I set out to find articles to send as gifts. Maybe a muff made from dog skin, or one of those carvings made from whalebone. The green stone carvings are expensive, as the stone must come from the middle island. And thank goodness the practice of mokomokai has just about stopped. The idea of sending shrivelled heads covered in tattoos is abhorrent. How did anyone ever think they would make intriguing keepsakes?

I'm also busy preparing clothing for the bairn. I'm larger now and find it more difficult to move about. Fortunately it is winter, so the garden doesn't require much work and I can sit inside by the fire with my sewing and embroidery. I've sent to Sydney for suitable material and consulted with Mrs White and Mrs Creed about sizes and styles. I've never paid much attention to such things, as I never expected to need that knowledge.

Having my first child at the age of thirty-one is rather terrifying. I remember stories of women wracked with pain and dying in childbed. What will happen to this bairn if I die, and it survives? Will one of the other missionary wives volunteer as wet-nurse? Or will Gideon have to seek out a native nurse?

And if I died, would Gideon remarry? I know I used to suspect him of proposing because I was the only eligible spinster here. It's still the case that there are very few pious women. I can't imagine he would take a native wife, as many do. Maybe he would follow William White's example and return home to find a wife. That worked well for William, although Eliza hasn't managed to tame him as was apparently hoped, just as I have been unable to soften Gideon's stubbornness.

I'm not even sure what to do with a bairn. How will I know when to feed it, or how to bathe it? I do appreciate the company of children when they are a little older and I can talk to them. Gideon's friend, James Buller, often brings his young son Walter from the Kaipara to visit us, and I enjoy the company of the Hobbs children. It is times like these when I miss my mother the most. I think about my cousins in Thirsk, where there are plenty of older women available to offer support and advice.

As my time draws near, I try not to worry about the future. My fate, and that of my bairn, are in the Lord's hands. I hope that the trials I have suffered already, and those I will endure, will secure my place in the next world.

'I'm out of inspiration today, Mary,' Gideon says. 'Can you help me?'

'With your sermon? How do you expect me to contribute?' I pause in my sewing as I look across at my husband.

'Well, what do you like to hear about? What's your favourite hymn, for instance?'

'I'll give you a clue,' I say, and I start to hum.

Gideon looks embarrassed. I already know he isn't very musical. He never takes the lead in singing, even when he is conducting a service.

'I'm sorry, I don't recognise it. Give me the lyrics, please.'

'Behold, fond youth, that busy bee,
How swift she flies from tree to tree
Extracting flowery sweets
Thus cheerful all the day she'll roam
At evening seek her much loved home
To treasure all she meets.'

'Of course,' Gideon says. 'How could I have not known that?'

'I especially like that it speaks of female endeavours,' I say. 'And then it also mocks the male.'

'Mocks us? In what way?'

'While yonder drone in sunny haunts

Who just supplies his present wants
Nor heeds the passing hours;
Soon bleak December's piercing air
Shall mock his want of timely care,
And chill his vital powers.'

'Another of Mr Charles Wesley's great works?' Gideon asks.

I nod, trying to restrain my grin.

'And you expect me to find a lesson in this?'

'It is speaking to make the most of what Providence bestows!' I tease.

'Thank you. But I think I will instead search the Holy Book for my inspiration. Maybe we can have that hymn, though, if you like it so much. But you'll have to lead the singing. I can't be seen to endorse female superiority, no matter what Mr Charles thought of it.'

CHAPTER 18

The birthing pangs come on suddenly, much worse than I expected. Fortunately, Gideon is home. He asks Hera to send for the doctor and our neighbour, Mrs Young. Beyond that, he is not much use. He kneels by our bed, praying, while I writhe in agony and Mrs Young mutters comforting nonsense and provides damp cloths and sips of water.

Then time disappears. The rest of the world is no longer real. All my attention is on the pain in my belly and back, trying to breathe between my groans. I don't care about the outcome. All I want is for it to finish. Even death would be preferable, although I know better than to utter this blasphemy to Gideon.

Finally, embarrassingly, Mrs Young is looking under my skirts and telling me to push. I don't think I have the energy to do so, but then my body takes control anyway. I feel a rush of movement, and the next moment Mrs Young is standing with a bloody mess in her hands.

She wipes its face and swaddles it in a wrap of cotton. It opens its mouth and screams.

'Mrs Young?' Gideon asks. 'The child? Is it unblemished?'

I want to tell him it is a ridiculous question. The Lord has given us this baby, and we have only to love it. God alone is faultless.

'A perfect little boy, praise God. Would you like to hold him while I attend to your wife?'

I watch Gideon gingerly accept the bundle, looking down at the tiny face in wonder. Mrs Young bustles about under my skirts again, but I take no notice as Gideon brings our child to me. I lift a hand to stroke the soft cheek, and the bairn opens his eyes – so blue! – and turns his pouting mouth towards me.

The anguish I have suffered over the past hours is nothing, forgotten in an instant. The reward of this bairn is worth every minute.

'A boy,' Gideon says, 'a son. Someone to carry on the lineage.'

I think about how my brother left no heirs. Now at least the Bumby blood will continue. I look at my husband, still entranced by his baby son's face. 'Do you think we could call him John?' I ask.

In front of Mrs Young, he kisses my cheek. 'Better than that,' he says, 'we shall call him John Bumby Smales. There is no one better for him to emulate than his late uncle.'

I find tears filling my eyes. 'You know, I can already feel Brother's devout spirit filling his soul.'

It is amazing to think that this new person has emerged into the world through my body. This morning, he was a thought, an anticipation, truthfully a burden. Now he is here, the centre of our world, a marvel of the Lord's creation.

Mrs Young takes the bairn from Gideon. 'Get your girl to give you a cup of tea,' she tells him. 'When you are finished that, you can bring a cup in for Mrs Smales.'

When he has left, she shows me how to suckle the bairn. Little John is eager but it takes some pushing and twisting for his mouth to find the right place. Mrs Young leaves us together for a while, then returns to give further instructions. The bairn and I would stay like this forever, I believe, but she warns against this.

Then she gives her farewells, promising to look in on us soon, and returns to her own house. I sink into an exhausted sleep, my eyelids lowering on the sight of my son sleeping peacefully beside me, where Mrs Young has placed him in the bassinet, hung from the ceiling.

When I wake, it is to see Gideon gazing adoringly at the bairn while the doctor examines him.

'Congratulations, Mr Smales. You have a fine healthy son. Now, if you would just leave us for a few minutes, I will examine your wife.'

His hands are cold and intrusive, but he pronounces himself satisfied with Mrs Young's work and leaves again.

The next three days are a mixture of elation, frustration, and fatigue. Little John is unpredictable. He'll wake after twenty minutes of sleeping, screaming as if in pain. I don't know whether he is hungry, uncomfortable, or truly in need of help. Gideon walks the room with him cradled over his shoulder, singing lullabies I've never heard. Other times, John sleeps for so many hours we are concerned he no longer breathes.

'I don't think I can do this,' I tell Gideon in the middle of the night. 'I'm tired and scared that I'll make a mistake.'

'You're doing fine,' he says. 'After what you went through, I'm not surprised that you're exhausted.'

Life begins to settle down, with us learning the rhythms of a newborn. Mrs White arrives for a week which frees Gideon to resume his work. She shows me more about the care of an infant – how to wrap him securely and place him in a crib under a tree to sleep. It is such a busy time, and I wonder how other mothers cope with both a new bairn and other children. I wish I'd provided more support to my friends and acquaintances during their time of new motherhood.

I think back to when Jane Hobbs was forced to move to Pakanae, shortly before Edward was born. There was no need for the family to go just then. Again, it was Mr

Hobbs enforcing his will on his family. How I resent his attitude. Instead of staying at Mangungu and having my support, Jane was dragged down to this much less hospitable place.

When I'm finally allowed to leave my bed, I find that Mrs White has improved the running of the house. The girls are more attentive and less surly. Corners that were previously neglected are now clean. Outside, the weather has become summery, the breezes warmer and more gentle. I gradually increase my walks around the area, young John in my arms wrapped in a shawl, showing him off to the interested aunties and kuia.

The threat to Gideon's life seems to have disappeared, and the local natives are once again more accepting of our presence. Nothing further has happened with Mr Hobbs, and we have relaxed back into our previous work. I still miss Mangungu, but life in Pakanae isn't so bad.

I am still in a daze of young motherhood when I notice the natives becoming restless. They are meeting and muttering, giving me sideways glances. Is this a resumption of the threat to Gideon? I wonder. He is away again on his circuit, not due back for several days. I wonder whether to send a message, asking that he return.

Eventually I ask Hera what is bothering her people.

'Maketu kill white lady,' Hera says.

I have many questions, all needing answers. Who is Maketu? And which white lady has he killed? Where has this taken place? And why are the local natives upset about it?

Gradually I tease the story out, and it is the stuff of my nightmares. A widow in the Bay of Islands, Mrs Roberton, was farming when she was attacked by a native farm worker and killed. This Maketu then proceeded to kill Mrs Roberton's young daughter, to throw her son from the cliffs onto rocks, and kill chief Rewa's three-year-old granddaughter. Mrs Roberton had been looking after the child for Rewa.

The natives reluctantly gave the killer up to the authorities and are now waiting to see what justice looks like under the Treaty. The young man was taken to Auckland for his trial.

'Maketu say they hurt his mana,' Hera explains.

I know that a native's honour, or mana, is a powerful force. It is easy to offend them and hard to apologise. Their way is to exact revenge, called utu.

I'm relieved when Gideon arrives home safely, and even more so as the grumblings around us subside. It seems that the natives are prepared to let English justice take its course.

Even so, it emphasises just how vulnerable we are here. Particularly in Pakanae, where we are isolated and at the mercy of the local tribe, who are much less friendly than Nene's people. I hope the work of civilizing them is taking hold, so that my bairn can grow up safe in this unpredictable country. The thought of a fate such as the one the Robertons suffered is too horrendous to imagine.

The red shawl of flowers which cover the pōhutukawa trees is even more stunning at Pakanae as Christmas approaches. I often surprise Hera, who can be found shirking her work and watching the bees as they come and go at the hives. It's hard to reprimand her interest.

The Wikitoria hive has recovered and calmed down. I'm unsure whether my trick of giving them new brood has worked, or whether they already had a new queen waiting to emerge. I know some beekeepers would find the current queen and remove her, so that the bees can make a stronger queen, now that they have more resources and there are a greater number of drones about. But I prefer to let the bees decide these matters for themselves, wherever possible. Usually, if the queen slows down her laying, the hive replaces her. I've even seen, one time back in England, a hive where there were two queens at once. Mother and daughter, Mr Webster suggested.

Hera finds another auntie to weave skeps. This auntie uses lengths of flax, scraped clean with a shell, but the baskets she produces are as sturdy as any I've seen. I test them by loading the top with a plank and stones, to see whether they'll collapse when the weight of a full load of honey drags on the flat tops. They merely settle more firmly onto their bases. This time I've got her to make swarming hives, without the aperture at the top. Rather than loading a duplet above the skep, I'll get the auntie to make an eke, a ring to put at the base, which raises the skep slightly higher and provides a little more space.

Hera delights in chasing swarms, which is lucky as I don't have the time or energy to do that. She grabs up a spoon or key and a pot, running out at the first inkling of a swarm about to leave. A child carrying the empty skep scuttles behind her. Then the 'tanging' begins, Hera loudly banging the metal utensils together. It is a warning to everyone that a swarm is imminent, and encourages the bees to settle in a nearby tree.

I struggle to find housegirls. I'd like to replace Hera, who is obviously more suited to working outside. But the new industry, preparing flax for rope to be shipped overseas, is enticing all the women away from domestic work. The men are paid well for the kauri timber they harvest from the forest. In Mangungu, it had been an honour to work at the mission in exchange for European clothing, food and the opportunity to learn to read and write. Here, those incentives don't seem as enticing.

I think back to my life at Mangungu. What did I do with all my time? Now, it seems that every waking moment is taken by attending to the bairn. No sooner have I bathed and fed him than I need to change him, burp and lay him down for sleep. Then there is all the washing, which isn't pleasant. No wonder I can't find housegirls.

Still, it is compensated for by moments of pure joy. Moments such as when John grips my finger, babbling away, and smiles before drawing my finger to his mouth. Those occasions when he sees Gideon and flaps his arms in excitement. If he is crying, held by one of the girls while I try to catch up with the mending, and then he sees me, he stops sobbing and instead starts hiccupping as his distress eases. Those are the times I feel especially blessed.

Gideon is always interested in the news that ships bring from other parts of the country. Lately, he's been hearing about the new settlements of Port Nicholson, Port Nelson and Canterbury. I tell him about the discussion I overheard between my brother and the Wakefields.

'I believe your brother was correct,' Gideon says. 'As I've been learning in my discussions with Nene and the other chiefs, they have quite a different concept of land ownership. I'll write to Mr Beecham in London and encourage him to make further submissions against the colonialism of New Zealand.'

A letter from John Whiteley urges Gideon to compromise in his stance with Mr Hobbs. At Mr Whiteley's prompting, Mr Waterhouse also sends a letter to Mr Hobbs, instructing another Special District meeting to be held to sort out the dispute.

'I apologised for not passing on the letters, and this time my apology was accepted,' Gideon tells me. 'I said I hadn't been aware that Mr Hobbs had been appointed my senior, and the Chairman. And then, would you believe it, Mr Hobbs also said that he had been at fault. So, it seems our enmity is resolved.'

However, only two weeks later at the Northern District Meeting in Mangungu, Gideon is caught unawares by an attack on his proficiency in the native language.

Gideon has been expecting a promotion from 'on trial' to 'full connexion', where he will be fully ordained in the Church. Instead, the Meeting expresses concerns that he has not been making diligent application in trying to learn the language and requires him to serve at least another year on trial.

'Mr Hobbs is singling me out, I'm sure. He's never liked me, and now he is making it impossible for me to stay and do my work.'

'But what else can we do?' I ask.

'I've been discussing it with Mr Whiteley. Now that he's in charge of the Southern District, he suggests that I move there.'

The Meeting is in favour of Gideon being given the station of Porirua, in the Southern District, but again Mr Hobbs has objected.

'Unless the Southern District can send a replacement north immediately, I cannot let Mr Smales leave the Newark station unattended.'

Gideon fumes the whole way down the river to Pakanae.

'I'm still at the mercy of Mr Hobbs, with no end in sight,' he tells me. 'I wish to leave, but he forces me to stay.'

'Have they no-one they could send in your place?' I ask.

'Can you imagine anyone volunteering to come into Mr Hobbs' district?'

'I see your point. And this area seems to be much poorer than when I first arrived. It is getting harder to make converts.'

'I feel we should pack up and move away, regardless of the instructions of the Meeting. The new settlements in Port Nicholson and Port Nelson surely need more preachers. We could try our luck there.'

'But is there anywhere to live?' I ask. 'I've spent some nights in a raupo hut when I first arrived, but never in winter, and not with a bairn.'

The smile of five-month-old John at that moment gives Gideon pause. It will be winter soon, and he has me and the bairn to consider.

'And I have the suspicion that I might be in the family way again,' I say. My nausea has returned and can't be ascribed to a boat journey. It is far too soon after John's birth for my comfort. Gideon has tried to abstain, I know, but we've both found it hard to forego our physical closeness.

'A brother or sister for John. That's wonderful news, Mary, although I don't wish the experience of birth on you again. You're right, we don't really have a choice about moving. We'll remain at Pakanae, and I'll work harder to improve my language proficiency. If only Mr Hobbs would ease off in his persecution.'

Hobart Town, Van Diemen's Land
7 April 1842

To All Our Loving Friends in New Zealand,
It is my saddest task to inform you that my husband, John Waterhouse, died last week here at home. He was only 52 years of age. He had caught a chill while out preaching and was already weakened from his recent voyage to Fejee.

We were all able to be with him at the end, and he knew his end was nigh. As he said, 'The righteous man hath hope in his death', and we believe it to be true. A more righteous man we never met.

Mr Turner will take over the role of General Superintendent until we have directions from the Society at home.

We send our love to you all and know that we have yours in return.

In sorrow, but also in the care of the Lord,
Mrs Jane Waterhouse

We are all distressed, but particularly me.

'He was my good friend, and like a second father to me,' I tell Gideon. How fortunate that he'd pressed me to marry. I can't imagine my life without my precious son, or indeed my husband.

'A wise and gentle man,' Gideon agrees. 'I do not know how we will fare without him.'

'He travelled with Brother and me on *James*,' I remind Gideon. 'If it weren't for him, I am not sure Brother would have come to New Zealand.' I wipe my eyes again. 'I must write to Mrs Waterhouse immediately.'

I've abandoned my teaching. Few students show any interest, and my time is now occupied by young John. Gideon spends more time travelling to outlying districts of his circuit, although he finds the natives stirred up by Te Wherowhero, who refused to sign the Treaty.

One day I answer the door and find Mr Hobbs there.

'Oh, I'm sorry. My husband is away on circuit,' I tell him.

'I know,' he says. 'May I come in?' He takes off his hat as he comes through the door.

I'm embarrassed that the house is untidy, with dishes still on the table and John's blocks of wood scattered on the floor. Luckily, John has just gone down for a nap.

'How can I help you?' I ask. I'm not going to offer him tea. I want him to state his business and leave.

I see him glance at my belly, starting to show the evidence of new life. It takes him by surprise, his mouth forming a brief 'O'. I feel a frisson of resentment, this man with his myriad of children judging us. But I try not to let it show.

'I've come to ask for your help with your husband,' he says.

I think I detect a smirk to his lips. I nod for him to continue.

'As you may know, he has not made sufficient progress in learning to speak the native language. The District believes that it is impeding progress in our efforts down here at Newark.'

I never call my home Newark. To me, it is the native Pakanae. How are we expected to learn the native tongue when we rename places?

'Not everyone has an aptitude for such things,' I tell him. 'However, I know that he is working to the utmost of his ability.'

'Are you sure?' Mr Hobbs says, impudently I feel.

'Of course,' I reply. 'You may be surprised to find just how far he has progressed. And I'm certainly assisting him where possible. Now, if that was all?' I want the man out of my house.

After he has left, I think about how he purposely chose to come at a time when Gideon was away. It is a long and arduous journey to get here from Mangungu for such a short discussion. What was he hoping to achieve? Maybe he was planning to sow dissent between us, for whatever evil reason he had. If so, his plan has backlashed. While I've always been supportive of Gideon, now I'm determined to do everything I can to assist him.

Or maybe that was Mr Hobbs' plan after all?

The winter drags by, lightened only by John learning to crawl. We are restricted to the house most days, and John wears out his clothing at the knees as he explores the floorboards and carpets of each room. On fine days I take him for walks around the mission, but it is usually too damp to allow him to crawl outside. There is a sandy beach for particularly nice weather, where he ingests shells and seaweed if I'm not watching closely enough.

Both Gideon and I are delighted with our child. Every day it seems that he learns something new, or gains another ability. Any visitor to our house is regaled with stories of how clever and how alert he is, until the proud parent finds themselves embarrassed at how they've monopolised the conversation.

Then John starts to walk around the furniture. I learn to keep my knitting up high after he tangles a whole skein when my back is turned. I often recover small items just before they find their way to his mouth. I begin to struggle with carrying him as the second child swells in my belly.

In September, Gideon travels down to Kawhia and preaches in native to an audience of about two hundred and fifty. The work we have been doing together pays off. John Whiteley observes and afterwards congratulates him on his progress.

'Mr Whiteley is a kind supporter,' Gideon tells me when he returns home. 'He says he will call for my instatement and relocation to Porirua.'

'But will Mr Hobbs let you go?' I ask. 'There is still no replacement for you here.'

'No, but I might go without his permission. Don't worry, though. I'll wait until the new baby arrives.'

Then comes the news that Governor Hobson has died, in his new capital, Auckland. Mr Willoughby Shortland, who accompanied the Governor to Mangungu twenty months ago, is now the temporary Governor.

'All these great men who are leaving us. What will we do without them?' I ask.

'It's the natural order, for the old to die and make room for the young,' Gideon says, rubbing my belly.

'But they take so much knowledge with them. And they weren't especially old. Mr Waterhouse was only 52, and Governor Hobson was a few weeks short of 50.'

I've just turned 32, and fifty no longer seems very far away. But Gideon is not yet 25, so perhaps that is why he shrugs his shoulders.

'We don't ever know the length of life allotted to us,' Gideon says. 'That's why I'm not prepared to remain here indefinitely.'

CHAPTER 19

John celebrates his first birthday, and eight weeks later his brother is born. Horatio Hewgill Smales arrives without fuss two weeks before Christmas, giving me just enough time to organise for the holiday.

'Look, darling boy, here's your new brother,' I say, as John is brought in to visit me. 'This is Horatio, after the great mariner. And Hewgill was my mother's maiden name, then my brother's middle name.'

Young John pokes him with a finger.

'No, don't do that. You must be gentle with him. Oh Gideon, can you believe John was ever this little? Or so helpless? They'll be wonderful playmates when they get older. How lucky we are to have two boys.'

'I admit I was rather hoping for a daughter,' Gideon says. 'Although boys do run in my family.'

'But how could you wish for Horatio to be anyone other than he is?' I ask, feeling absurdly defensive of my new son.

'No, of course I don't. He's perfect just the way he is, and you're my very clever wife. Now, I'll take this big boy away and let you rest.'

'Mama, Mama,' John protests as he is carried from the room. Then I hear him giggling. Perhaps Gideon is tickling John to distract him.

Only yesterday John seemed still a bairn, but now he is walking, talking, and is an older brother. I snuggle back into the pillows. I am content. I have my husband, my two sons, and all is right with the world.

Meanwhile, Gideon spends his time trying to leave the Hokianga. He starts looking for a ship that can take us south. *Triton*, which would have been the obvious choice, is in Hobart Town and likely to be going to the Islands before coming on to New Zealand.

'I think we should pack most of our belongings, ready to depart,' he tells me, a few weeks after Christmas. 'If we can't get a ship, maybe we should try an overland route.'

'I'm not sure I can manage that,' I say. 'I've only just recovered, and Horatio is still so young.' I haven't left the Hokianga area in the four years I've been in New Zealand. I remember my brother's exhaustion in our early days whenever he travelled from home. I know Gideon is strong and able, but I'm no Lady Franklin.

'I don't feel I can accept the rule of Mr Hobbs any longer,' Gideon says. 'Any other place would be preferable.'

'Please,' I say. 'Give me a little more time with the bairn. He's not as strong as John was.'

I feel as if I am being pulled between my husband and Mr Hobbs. I can certainly understand why Gideon finds it so difficult to work under that man, but he is not inclined to compromise either. Of course, I knew this when I married him. Mr Waterhouse made a point of warning me. But what can I do? I respect that Gideon has principles, and I know that he is working hard to bring the natives to our Lord. If only that horrible Mr Hobbs would leave him alone, we could be happy here.

I'm sitting on the spit of sand that reaches into the river, trying to feed my new bairn. Horatio is fussy, turning from the breast and often drawing his legs up and screaming. I'd thought that having a second bairn would be so easy, a repetition of all that I'd done for John. But

I can see that Horatio truly is a different child, and what worked for one may not work for the other.

John is paddling in the waves. He's devised a game where he ventures close, then runs shrieking as the wavelets get near his feet. I'm keeping an eye on him. The ripples are gentle and lapping today, and the sun is fierce. If he tumbles into the water, it won't take long to dry his clothes.

I take a moment from gazing at my lovely boys to look across the water at the far shore. The dirty turquoise of the river gives way to the dark seaweed-green of the trees, rising steeply until they reach the cornflower-blue of the sky. There's not a cloud and only a sniff of a breeze. If I turn, I know I'll see the pōhutukawa in its last stages of dripping red. The birds have quieted in the heat and the noisy chirrup of cicadas have taken their place.

We've been here at Pakanae for nigh on two years, and I've come to love it almost as much as Mangungu. Here, I get to taste the salt on my skin when the wind blows in from the Heads. The smell is changed, without the muddiness of the mangroves. It is more primal, colder and more windswept. I would have thought I'd dislike being close to the ocean and reminded of its vastness and power, after my journey to this country.

Yet it is here that I've found peace. My babies and my husband have given me a life back. I know that I should be content to serve the Lord's purpose, but it is easier to care for people, especially those you love. And I am doing His work, by supporting Gideon.

I don't want to leave the Hokianga, though I know I must. I'm savouring every moment, committing it to my memory, so that I can tell the boys where they came from when they are grown.

Three years to the day after the first signing of the Treaty, the brig *Guide* arrives in the Hokianga. She is relatively old and I'm worried that she's unsafe, but Gideon knows her captain, one of the Wesleyan agents in Sydney.

'It's fortuitous,' he says. 'This is our chance to escape.'

'But where will we go?'

'We have friends. Mr Whiteley in Kawhia. Mr Wallis in Whaingaroa, or Mr Aldred in Port Nicholson. New Zealand is a bigger place than it was when we arrived, Mary. There are opportunities for us elsewhere, and none here.'

He books a passage south, loading most of our belongings on board and storing the bulkier items with friends.

I call for Hera.

'We are leaving,' I say. 'Ke te wehe atu. You will have my bees. You are good with them. Look after them, treat them well. Tiaki nga pi.'

I cannot bear to leave without visiting them, my precious bees. I put Hera in charge of the children and go out to the hives. They deserve to be told that they are changing owners. As I approach, I can see that one of the skeps is looking very strange; much darker than usual, and pulsating. I get closer and see bees covering the skep like a blanket. I stand in front of them and they rise into the air, flying past me as a wave of black bodies takes to the sky. They form a spiral of communing bees, beautiful against the blue background. It fills me with awe at the Lord's wondrous creation, this group of simple creatures that can work as one being.

They must already know that changes are coming. This is their farewell to me. It is very late in the season for them to swarm, and I am concerned. Hera is busy looking after my sons and can't follow them into the forest. The bees won't have much time to gather stores and build another nest before winter arrives. I fear for their survival.

Then, as I stand fretting, they descend to the skep and a stream of bee bodies clambers back into their nest. I am immensely relieved. It has been a demonstration of their awareness of my departure and their loss.

How I would like to take them with me. They feel a part of me. Likely none of them, not even the queen, are the ones who travelled out to this country with me, but their ancestors were on that voyage. This expedition is different, though. I have many more obligations, less time to ensure their safety, and my future is even less certain. They will have a better chance of surviving if I leave them here.

'Farewell, my beauties,' I say, wiping my tears. 'May you flourish here, in ways I have been unable to achieve.'

I walk away from my hives for the last time.

The children and I are taken out to the vessel and installed in a small cabin. It reminds me of my voyage on *James*, but this ship is even smaller and less well accommodated. I'm not looking forward to the two weeks it will take us to reach Port Nicholson. I worry that young John will be injured or fall overboard, and I'm unable to imagine how I'll keep him entertained for the whole journey. Not when I also have Horatio to care for.

Gideon doesn't inform Mr Hobbs, fearing that he'll find a way to interfere, but instead leaves him a note. At the last minute, he loads some of the mission's goats and chickens on board.

'Don't they belong to the mission?' I ask.

'We've raised them. And they'll be useful as gifts for the Wesleyan stations we're about to visit.'

We stand on deck as the ship slips quickly down the river, waving to our native friends.

'I feel wretched that I am deserting them,' Gideon says. 'I only hope that another missionary can take my place in the fight for their souls.'

Across the water come shouts of farewell, 'Haere atu ra, haere ki runga ki toa kāinga.' They are sending us off to our new home with their best wishes. I wipe my eyes, blow my nose.

I'm leaving home again. I've spent the last four years here, growing to love the people and the countryside. So much has happened to me on the shores of the Hokianga, and now I might never see it again. We're heading off into the unknown, back on board a constricted ship, this time with two small children.

And I realise, for the first time, that this country is truly my home. While I might still call England home, I don't think of it as where I belong. This is the birthplace of my children and where I can do the most good, alongside my husband. I am not trapped here but choose to stay. I just can't remain in the Hokianga, unfortunately.

I strain to see my bee skeps one last time, so unwillingly left behind. How will they cope without me? Will the bees swarm into the forest and establish themselves in the hollows of trees? Does Hera know enough yet to care for them? And what of my other hives, at Mangungu?

The Hokianga and my bees are the last links I have to my brother. This is my final farewell to him. And yet, no. Down below in the cabin are my sons, John Bumby Smales and Horatio Hewgill Smales. Both named for my brother.

I'm not leaving him at all. Just journeying onward, continuing his work. With a final wave, I turn and head to the cabin. And the future.

AFTERWORD

I never thought I'd see this day, but here I am back on a ship, heading home to England. I have no bees with me this time, but it feels as if I'm the queen bee, the way my husband and daughters hover around me. Sadly, even though SS *Northumberland* is a modern steam ship, I am suffering even more than on my outward journey, and I seriously doubt that I will see the shores of my old country again. The journey is rough with high seas and squalls, and I am confined to bed.

A lot has happened in the nineteen years since we left the Hokianga. These lovely daughters, for a start. Polly, Rosie, Fish (that's what we call her, although her real name is Felicia), and Sophia my baby, not quite ten years old.

We're travelling to visit our son and the girls' brother, John. He's been living in England, studying history at London University. His time at the new Wesleyan School in Auckland must have been of some worth. Unfortunately, little Horatio never did well after we fled the Hokianga on that cold, wet ship, and he died shortly before his first birthday. Then, several years ago we lost our third son, Giddey, when he was thrown from his pony. He was only eleven and I miss him sorely.

After leaving the Hokianga, Gideon and I spent most of that year in the new settlements of Nelson, Wellington and Porirua, before settling in Aotea Harbour, near the Whiteleys at Kawhia. There, we established Beechamdale, named after John Beecham from the Wesleyan Society in London. The land was fertile, and we created a farm, growing wheat, oats and barley and raising animals, aiming to become self-sufficient and not relying on donations from abroad. Gideon even helped some of the local people to build the country's first water-driven flour mill.

Gideon, concerned for the futures of our children, started acquiring land on the outskirts of Auckland, not far from the fencible village of Howick. We supported the establishment of Wesley School and, as well as John, native men from Aotea attended, making us proud of our contribution to their early education.

Our family left Aotea and moved to Auckland six years ago, after a long-standing conflict within the Wesleyan Missionary Society. Gideon decided to concentrate on being a farmer, although his faith remains unshaken. He has even had a lovely stone church built on our land, open to both Wesleyan and Anglican ministers. The timing of our move was fortuitous, as there is serious unrest happening in Taranaki and Waikato. I fear for our faithful New Zealanders. We thought we had shown them how to achieve peace.

I can hear a buzzing in my ears now. I remember my bees swarming in their farewell salute as I left them for that last time. I wish I'd had the strength and energy to take them with me. One of my daughters might have been able to take over their care.

I must remember to ask my girls to visit the bees when I go, to shroud them in black cloth and tell them that their keeper has departed.

I will see them again in that better place that is promised to us.

HISTORICAL NOTES

I have tried to keep to the facts, as known, about the many historical characters that appear in this book. Any mistakes are my own, and I apologise for them.

I have kept place names as they were at the time. The Hokianga was called a river, not a harbour. Raglan was named Whaingaroa, and Russell was Kororareka.

The native people of the land were known as New Zealanders, not Māori. As this term is unwieldy, I have usually called them natives, and occasionally savages, which is what some people of the time would have said.

Australia, and especially New South Wales, were referred to as the colonies. Tasmania was still known as Van Diemen's Land. Tonga was called The Friendly Isles.

However, this is a novel. Much is invented, as I have no way of knowing what Mary and her companions thought and said and did. Mary's diary is sparse and only partly helpful.

Both Emma and Marianne Hobbs did end up marrying missionaries.

The New Zealand beekeeping industry is now large and successful, especially with the rise of mānuka honey. And the bees we keep today are different stock to Mary's. The invention of the Langstroth box hive has changed many aspects of beekeeping. It is no longer permissible to keep bees in skeps of any kind, as it is a legal requirement that all honeycomb should be able to be taken out and inspected for disease. I am grateful to the Bees for Development Trust in the UK for their course on making straw skeps, and for the books they have published on skepping. There is still much to be learnt from the old ways.

Mary did not reach England and was buried at sea. She was only 51.

GLOSSARY

Atua – Māori gods and spirits
Barm – live yeast for breadmaking
Bole – recess in a wall to house a straw beehive
Brood – larval stage of a bee's development
Chafing dish – vessel used for holding burning charcoal or other fuel
Cloam – cow dung mixed with sand or ashes
Drone – male bee
Duplet – empty skep added on top of a skep
Eke – short cylinder added below to give a skep more height
Festoon – where the bees hang together, leg to leg, in lace-like sheets between combs
Heft – to judge weight by lifting
Hogshead barrels – 54-gallon wooden vessel with one end open
Humblebee – old English word for bumblebee
Kāinga – Māori village or settlement
Lucifers – friction matches made of phosphorus
Poop deck – the roof of a cabin built at the rear of a ship
Sling – method of collecting honey without killing the colony
Skep – basket made of straw or reeds, used for beehives
Spleet – small strip of split wood
Split – halving a colony of bees to make two hives
Storeyfing – adding a further skep on top of a hive to increase room

Swarm – group of bees leaving their home hive to start another
Tanging – banging metal objects to encourage a swarm to settle

ACKNOWLEDGEMENTS

My greatest thanks are to my husband Peter, who traipsed around England with me on my research, encouraged me throughout my master's programme, and accompanied me around the Hokianga.

I'd also like to thank fellow beekeepers Peter Biland (who first mentioned Mary's story to me), Chris Park, Dr Graham Dyche, and the Franklin Beekeepers Club.

I owe much to David Bumby (distant cousin of Mary) and Anne Middleditch, John Steele, and Peter Barrett for their excellent books on Mary, John and Gideon, and their encouragement.

My thanks also to writers John Cranna, Lucille Henry and Tatjana Walters, Graeme Lay, Siobhan Harvey and Janet Pates for their input. The Historical Novel Society Australasia (HNSA) conferences helped inform my writing.

I appreciate the work of my early reviewers, Ruby Rutherford, Mary Amore, Tricia Carter, Lynne and Terry Wall, Wikitoria Makiha and Tanya Batt. You have helped me steer clear of some awkward mistakes!

Thank goodness for the organisations which preserve our history. I especially appreciate the Hocken Library, the Mangungu Mission, the Hokianga Museum and Archives Centre, the Auckland War Memorial Museum, and the helpful people of Thirsk and Northallerton.

SOURCES

Barrett, A. (1853). *The Life of the Rev. John Hewgill Bumby.* Manchester.

Barrett, P. (2022). *Mary Bumby's Bees, 1839-1840; myth, fact, mystery.* Caloundra, Australia.

Middleditch, A. and Bumby, D.F. (2018). *Mary Bumby; The first person to take honeybees to New Zealand.* Northern Bee Books.

Steele, J. (2011). *Smales' Trail.* Bleak House.

Williment, T.M.I. (1985). *John Hobbs* (1800-1883). VR Ward, Government Printer

www.ingramcontent.com/pod-product-compliance
Lightning Source LLC
Chambersburg PA
CBHW071703160426
43195CB00012B/1558